Historical Animal Geographies

Arguing that historical analysis is an important, yet heretofore largely underexplored dimension of scholarship in animal geographies, this book seeks to define historical animal geography as the exploration of how spatially situated human–animal relations have changed through time. This volume centers on the changing relationships among people, animals, and the landscapes they inhabit, taking a spatio-temporal approach to animal studies. Foregrounding the assertion that geography matters as much as history in terms of how humans relate to animals, this collection offers unique insight into the lives of animals past, how interrelationships were co-constructed amongst and between animals and humans, and how nonhuman actors came to make their own worlds. This collection of chapters explores the rich value of work at the contact points between three sub-disciplines, demonstrating how geographical analyses enrich work in historical animal studies, that historical work is important to animal geography, and that recognition of animals as actors can further enrich historical geographic research.

Sharon Wilcox is the Associate Director for the Center for Culture, History, and Environment in the Nelson Institute for Environmental Studies at the University of Wisconsin-Madison. Her research explores the ways in which conceptions of place and value are constructed for terrestrial mammalian predator species in historical and contemporary contexts. She is the author of the forthcoming monograph, *Jaguars of Empire: Natural History in the New World*.

Stephanie Rutherford is an Associate Professor in the School of the Environment at Trent University in Canada. Her research inhabits the intersections among the environmental humanities, animal geography, and posthumanism. She is currently writing a book on the history of wolves in Canada. She is also the author of *Governing the Wild: Ecotours of Power* and co-editor (with Jocelyn Thorpe and L. Anders Sandberg) of *Methodological Challenges in Nature-Culture and Environmental History Research* (Routledge, 2016).

Routledge Human–Animal Studies Series

Series edited by Henry Buller

Professor of Geography, University of Exeter, UK

The new *Routledge Human–Animal Studies Series* offers a much-needed forum for original, innovative and cutting-edge research and analysis to explore human–animal relations across the social sciences and humanities. Titles within the series are empirically and/or theoretically informed and explore a range of dynamic, captivating and highly relevant topics, drawing across the humanities and social sciences in an avowedly interdisciplinary perspective. This series will encourage new theoretical perspectives and highlight ground-breaking research that reflects the dynamism and vibrancy of current animal studies. The series is aimed at upper-level undergraduates, researchers and research students as well as academics and policy-makers across a wide range of social science and humanities disciplines.

For a full list of titles in this series, please visit: www.routledge.com/Routledge-Human-Animal-Studies-Series/book-series/RASS

Historical Animal Geographies

Edited by Sharon Wilcox and
Stephanie Rutherford

Routledge
Taylor & Francis Group

LONDON AND NEW YORK

First published 2018
by Routledge
2 Park Square, Milton Park, Abingdon, Oxon OX14 4RN

and by Routledge
711 Third Avenue, New York, NY 10017

Routledge is an imprint of the Taylor & Francis Group, an informa business

British Library Cataloguing-in-Publication Data
A catalogue record for this book is available from the British Library

Library of Congress Cataloging-in-Publication Data
A catalog record for this book has been requested

ISBN: 978-1-138-70117-5 (hbk)
ISBN: 978-1-315-20420-8 (ebk)

Typeset in Times New Roman
by Apex CoVantage, LLC

Contents

Illustrations

Figures

Boxes

Contributors

Philip Howell is Senior Lecturer in the Department of Geography, University of Cambridge. He is the author of *Geographies of Regulation: Policing Prostitution in Britain and the Empire* (Cambridge University Press, 2009) and *At Home and Astray: The Domestic Dog in Victorian Britain* (University of Virginia Press, 2015), and co-editor of *Animal History in the Modern City: Exploring Liminality* (Bloomsbury, 2018) and *The Routledge Companion to Animal-Human History* (Routledge, 2018).

David Lambert is Professor of History at the University of Warwick. His research is concerned with slavery and empire in the eighteenth and nineteenth centuries, focusing on the Caribbean and its place in the wider (British Atlantic) world. He is the author of *Mastering the Niger* (University of Chicago Press, 2013), *White Creole Culture, Politics and Identity during the Age of Abolition* (Cambridge University Press, 2005), and co-editor of *Colonial Lives Across the British Empire* (Cambridge University Press, 2006).

Teresa Lloro-Bidart is Assistant Professor in the Liberal Studies Department at California State Polytechnic University, Pomona, specializing in environmental studies and sciences. She is a political ecologist working at the intersections of animal studies and environmental/science education. She uses multispecies ethnographic methods and historical archive research to understand how educational spaces and processes are inherently political and produce human–animal and human–nature relations within these political frameworks. Theoretically, she grounds her work in feminist, posthumanist, and biopolitical thought.

Ian MacLachlan is Professor Emeritus at the University of Lethbridge in Alberta with broad interests in economic geography and regional planning. In recent years, he has specialized in livestock and meat processing issues, regional economic models, and urban planning research in China. A Montreal native and Lethbridge booster since 1989, he retired in 2015 and now divides his time between the University of Lethbridge and Peking University Shenzhen Graduate School.

Jennifer Mateer is a Postdoctoral Scholar at the University of Winnipeg in the department of Environmental Studies and Sciences. Her work covers urban, rural, more-than-human, and feminist political ecology, with a current project focused on the impacts that new water management strategies have on the more-than-human world and the hydro-social cycle in India. Previously, she conducted research in Rwanda on topics related to geographies of health.

Heidi J. Nast is Professor of International Studies at DePaul University. Her interests lie in theorizing ontologies and epistemologies of fertility across cultural and political economic domains. Her work traverses: royal contexts in West Africa where concubinage was used to create royal children whose marriages forged territorial alliances; the industrial Machine as a sexed competitive device for displacing maternal ontologies source of value; pet humanization, maternal alienation and capitalist crises of consumption; and the Machinic end of the biological maternal.

Dominik Ohrem is Lecturer and Doctoral Candidate in the North American Studies program at the University of Cologne, Germany. His research and teaching interests include U.S. history, environmental history, animal studies, gender studies, and feminist philosophy. He is currently working on his research project about animality and human–animal relations in the context of American westward expansion. He is editor of *American Beasts: Perspectives on Animals, Animality and U.S. Culture, 1776–1920* (Neofelis, 2017), as well as co-editor (with Roman Bartosch) of *Beyond the Human-Animal Divide: Creaturely Lives in Literature and Culture* (Palgrave Macmillan, 2017) and The final title of the volume is (going to be): *Exploring Animal Encounters: Philosophical, Cultural, and Historical Perspectives* (with Matthew Calarco, Palgrave Macmillan, 2018).

Chris Philo is Professor of Geography at the Department of Geographical and Earth Sciences at the University of Glasgow. He has been at the forefront on animal geographies and co-edited the book *Animal Spaces, Beastly Places* (Routledge, 2000).

Harriet Ritvo is the Arthur J. Conner Professor of History at the Massachusetts Institute of Technology. She is the author of *The Dawn of Green: Manchester, Thirlmere, and Modern Environmentalism* (Chicago University Press, 2009), *The Platypus and the Mermaid, and Other Figments of the Classifying Imagination* (Harvard University Press, 1997), *The Animal Estate: The English and Other Creatures in the Victorian Age* (Harvard University Press, 1987), and *Noble Cows and Hybrid Zebras: Essays on Animals and History* (University of Virginia Press, 2010).

Camilla Royle is a Ph.D. candidate in geography at King's College London and is deputy editor of *International Socialism*.

Julie Urbanik holds a Ph.D. in Geography from Clark University and spent ten years as an award-winning educator and author. She is the executive director of The Coordinates Society, a nonprofit with a mission to cultivate curiosity and compassion through geography, and an expert geo-narrative media consultant for public defenders. She is the co-founder of the Animal Geography Specialty Group of the Association of American Geographers, co-editor of *Humans and Animals: A Geography of Coexistence* (ABC-CLIO, 2017), and author of *Placing Animals: An Introduction to the Geography of Human-Animal Relations* (Rowman & Littlefield, 2012).

Thomas Webb is a Postdoctoral Research Associate in the Department of History at the University of Liverpool. His dissertation examined World War II's impact on human–animal relations in Britain with a particular focus on food production and fighting. His research draws on interdisciplinary influences from social and cultural history, cultural geography, animal studies, and environmental histories of war.

Jennifer Wolch is Dean of the College of Environmental Design at the University of California, Berkeley. Her work in the area of animal–society relations has focused on the broad thematic of cultural animal geographies, and in particular interactions between people and animals in urban contexts. This research has focused on trans-species urban theory, empirical analyses of attitudes toward animals, community/metropolitan planning that is inclusive of animals, and critical assessment of animal-oriented architecture and urban design. With Jody Emel, she co-edited *Animal Geographies: Place, Politics and Identity in the Nature-Culture Borderlands* (Verso, 1998).

Acknowledgments

We would like to thank our editor, Faye Leerink, who approached us at the Association of American Geographers meeting in 2015 to invite us to submit the book prospectus that developed into this volume. The two editorial assistants who served on this book project, Priscilla Corbett and Ruth Anderson, have been invaluable in moving this volume to completion. Thanks are also due to Henry Buller, the editor of the Human–Animal Studies series at Routledge, for his support of this collection. We were lucky to have three very helpful anonymous reviewers, who pushed us to be more precise in our intent and framing of the book, leading to what we hope is a much better volume. Paul Robbins generously agreed to write a foreword for this book, and we are deeply grateful for his engagement with this project. Lastly, we would like to thank our contributors, who so thoughtfully engaged with the questions we raised around human–animal relations and how they evolve through time.

Foreword

In his book *Duino Elegies*, the poet Rainer Maria Rilke wrote: "and the animals already know by instinct we're not comfortably at home in our translated world" (Rilke 1977). I have puzzled over this passage for a long time. In what way is our world translated? What makes us so ill at ease there? And what would the animals tell us if they could?

By re-reading history through animals, or perhaps by writing animals back into history, I sense an answer to these questions may be at hand, and in it may lay a path through the murkiness of our social and environmental future. This is because animals have always travelled alongside humanity, creating and re-creating the Earth with us, responding to our meanderings and depredations, and setting new terms with which people have always had to contend. If animal trajectories both follow and lead ours, then tracing their path can better show us who and what we are, and maybe even provide us a glimpse of what we might become.

Consider Jacob Shell's (2015) remarkable observations in his *Transportation and Revolt*, a book that shows how animals – from sled dogs and pigeons to camels and mules – became companions and allies for revolutionaries, guerrillas, and indigenous communities seeking autonomy from state and empire. These animals, as a result, became targets of brutal force. This remarkable and repeated fact of history sheds some light on the geographic problems of governance, since it is the mobility of these companions that tests the limits of state power and provokes such aggression.

More than this, the very mobility of these animals and human lifeways may be key to adaptation and adjustment in a world transformed by massive environmental change. As spatial shifts and uncertainties in ecology, weather, and climate continue to mount, we might come to embrace the historically marginalized companionships of abject and insurgent people and animals, in order to cope, or even thrive. In this way, a history of animals and people also provides clues about surviving the Anthropocene.

This is merely one instance of how the traces of animals in history shed light onto geography, power, and humanity. Once the lid is off this line of inquiry, a Pandora's box of questions and contexts explode into view. The histories and geographies collected here by Wilcox and Rutherford are filled with these questions and contexts, across homes, cities, nations, and globe-spanning networks. By wisely bringing them together in one place, they have provided a roadmap for future thinking – and perhaps the key to unlocking Rilke's profound more-than-human mystery.

Paul Robbins
Director of the Nelson Institute for Environmental Studies
University of Wisconsin-Madison

References

Rilke, R. M. 1977. *Duino Elegies and the Sonnets to Orpheus*. New York: Houghton Mifflin.

Shell, J. 2015. *Transportation and Revolt: Pigeons, Mules, Canals and the Vanishing Geographies of Subversive Mobility*. Boston: MIT Press.

1 Introduction

A meeting place

Stephanie Rutherford and Sharon Wilcox

This book began, like many do, in a windowless conference room in the basement of a cavernous hotel. In October 2014, Sharon Wilcox circulated a call for papers for the forthcoming American Association of Geographers (AAG) annual meeting, inviting contributions to a session on animal histories. Stephanie Rutherford, looking for just such a session, responded to this call. They ended up organizing two sessions, bringing together a myriad of excellent papers that grappled with the question of how we explore the complex interactions and connections between animals and humans in the past.

This volume also began, to some degree, as a meeting place – a niche, if you will pardon the pun – of shared intellectual pursuit. Although there had been exceptional and groundbreaking work in animal geographies using a contemporary frame (Barua, 2016, 2017; Buller, 2014, 2015; Collard, 2012, 2014; Gillespie & Collard, 2015; Lorimer, 2015; Philo & Wilbert, 2000; Urbanik, 2012; Wolch & Emel, 1998), little of it was overtly historical. By contrast, environmental histories of animals (Coleman, 2004; Landry, 2008; Datson & Mitman, 2005; Pearson, 2013; Ritvo, 2010; Rothfels, 2002, 2008; Swart, 2010) were a real source of insight for our respective research programs, but they often did not take into account pivotal geographical concepts like scale and spatiality alongside agency, contingency, and change through time. As a result, we both had the impression that we were working in our own silos – Sharon on large cats and Stephanie on wolves and wolf hybrids – at the interstices among history, geography, and animal studies. As such, it was nice to find a fellow traveler with whom to compare notes about doing this kind of transdisciplinary research.

Even while our research and writing has haunted the edges between various disciplines, we are both firmly animal geographers. For us this means taking seriously the agency of the more-than-human world and acknowledging the capacity of all the earth's inhabitants to affect and be affected. As such, we reject the idea of human exceptionalism and, along with other scholars who embrace the posthuman turn, seek to unsettle the presumed boundaries between the organisms that make up this world. Indeed, humans and other animals are always co-constituted – our ontologies are relational, even if the effects of co-constitution are always asymmetrical. What this suggests is that we were never quite human in the way some of us thought we were, and that we need to be more capacious in our understanding of agency, decentering the human subject as the only beings capable of shaping the world. With a more generous articulation of agential capacity, all meetings become "contact zones" of "becoming with" (Haraway, 2008, p. 244), which are always improvisational, contingent, precarious, and transformative for all the agents doing the relating. As van Dooren, Kirksey, and Münster (2016, p. 1) maintain, "all living beings emerge from and make their lives within multispecies communities . . . situated within deep, entangled histories," with each irrevocably shaped by the encounter.

We were interested in exploring how the kind of relationality that felt easier (though still fraught) to elaborate in the present moment might be explored through time. How might we explore the entangled histories to which Thom van Dooren, Eben Kirksey, and Ursula Münster refer? Figuring out how one might access animal lives in the past – through which theoretical perspectives, disciplinary approaches, and methodologies – was the vexing question with which we both grappled. As many historians have noted, animals leave little record of their existence, as they do not maintain archives and histories where we might go in search of their past. John Coleman (2004) suggests that through genetic legacy, caring for their offspring, and other ways of communicating (e.g., scrapes, marking, vocalization), histories and memories are passed down through generations in ways humans cannot fathom. However, for scholars attempting to reconstruct lives of animals past, little remains that testifies to these prior existences. Individual lives are erased by chemical, biological, and physical processes acting upon the landscape that wipe away identifiable traces, tracks, and remains. Indeed, uncovering animals in the past frequently requires examination of sources that frame human-animal encounters as, what Michael Woods termed, "ghostly representations" that "speak on behalf of the animal" (2000, p. 199). Traditionally, historical research has often rendered animals into an absent-presence, spectral figures at the edges of historical change because of this lack of physical trace or capacity to intervene in the archive.

Recent scholarship has sought to unsettle this attachment to the archive, or at least its veneration as the source of all insight about the past. For instance, Erica Fudge (2017) has suggested that past subjectivities – human and nonhuman – have always been difficult to access; the archive is always already fraught. Indeed, she offers the corrective that both humans and nonhumans are often shaped by an archive not of their own making, so we should not imagine that narrating the lives of humans is a straightforward endeavor either. This is something certainly echoed by people of color, women, Indigenous people and others who have been written into or out of archives in ways that did not reflect their life experience, but did secure asymmetrical relations of power. Erica Fudge, along with other environmental history scholars, has emphasized the importance of reading historical texts "against the grain" (2017, p. 264), while broadening our understanding of "text" to include things like archaeological evidence and ethological data. These interventions, along with a more expansive notion of agency and a broader vision of the multiplicity of actors that make up our world, have been invaluable to geographers working at the intersections of animal studies and historical analysis.

The argument

The central argument of this book is that historical analysis is important to animal geographies; put differently, it opens up the possibility that those of us working in the area of animal geographies might do well to take a temporal as well as spatial approach. Citing Ann Norton Greene, Susan Nance contends, "historians routinely forget or refuse to see that 'animals change over time' and hence require historicizing just like humans" (2015, p. 6). The contributors to this volume aim to do precisely that: think about how the ways in which human and nonhuman agencies have shaped each other through time.

Ultimately, the editors of this volume remain geographers. So, while acknowledging the importance of this history, we also want to suggest that space, place, landscape, and scale are also essential to re-constructing animal lives of the past. As such, this book aims to begin sketching out a subfield at the intersections of historical and animal geographies. We seek to define historical animal geography as the exploration of how spatially situated human–animal relations have changed through time. More specifically, this volume is about the changing relationships among people, animals, and the landscapes they inhabit through time, taking a spatio-temporal approach to animal studies. Foregrounding the assertion that geography matters as much as history in terms of how we relate to animals, these collected chapters offer unique insight into what life conditions animals have encountered, how interrelationships were co-constructed, and how nonhuman actors can make their own worlds. Ultimately, the contributions found in *Historical Animal Geographies* explore the rich, yet largely unexplored value of the contact points between three sub-disciplines, demonstrating how geographical analyses enrich work in historical animal studies, that historical work is important to animal geography, and that further examination is needed of animals as actors in historical geographic research.

Historical Animal Geographies at a glance

This volume features fourteen diverse chapters from both noted and emerging scholars, representing the forefront of a spatio-temporal approach to writing animals into their own lives and histories. These chapters offer a broad range of contributions – epistemological and ontological, theoretical and empirical – as these scholars explore the unique methodological challenges of articulating an animal-centered historical geography. Broadly, these chapters engage with different ways of exploring how animals might be conceived of as historical agents in their relationships with humans. The chapters in this book explore socio-ecological histories of particular animals offering historical accounts of animal agency and subjectivity, taking seriously the notion that humans and animals have shaped each other in particular historical and geographical contexts. Each contribution explores the entwined co-beings of animals and humans – collaborative, conscripted, or coerced – across a range of contexts, from megafauna to microscopic, rural to urban landscapes, from animal lives lived in close habitation with humans to those most remote. In so doing, the chapters in this volume offer a thoughtful and inter-disciplinary set of approaches to considering the lives of animals and the ways in which they might be resurrected to illuminate new dimensions of the known and unexplored past.

Part I: The home – shared spaces of cohabitation

What does it mean to share a life? How do recognition and misrecognition impact how humans and nonhumans encounter each other in intimate space? In what ways are co-creation and collaboration reinforced or rejected? And how might animals make homes for themselves through time completely outside of human action or knowledge? The three chapters in this section grapple with these questions in fascinating ways. Philip Howell's contribution is animated by the provocative question: when did pets become animals? Underneath this seemingly straightfor-ward question rests complicated, intimate interspecies relations, shot through with ideas of home, notions of property, and the intermingling of love and domination. For Teresa Lloro-Bidart, "home" is extended to the institutional facilities that housed disabled Union veterans of the American Civil War. In these sites, nontra-ditional companion animals were deployed as a civilizing technology to inculcate white middle-class values in veterans who were damaged by the violence of war. In this way, the more-than-human became a tool for moral uplift within systemized functions and spaces of the home. The section ends with Camilla Royle's intrigu-ing work on the "small agencies" of earthworms. Using biology as a source for the construction of animal pasts, Royle explores the role of centuries-old worms in construction of the niches – or homes – that today's worms navigate. In so doing, Royle upends the suggestion that only people create history, with worms authoring lifeworlds for themselves that are occluded from human view.

Part II: The city – historical animals in and out of sight

Although we might invite pets into our homes, there are many animals with whom we unknowingly share space. For instance, until recently, nonhuman life has often been written out of urban histories and theories of urban space. In 1996, Jennifer Wolch wrote "Zoöpolis," a now-classic article taking urban theory to task for its implicit human exceptionalism. Wolch suggested that the city was teeming with nonhumans and that, by ignoring its liveliness, scholars risk not only doing vio-lence to nonhumans but also other struggles around space, place, and justice in the city. This influential piece is reprinted as Chapter 5 here. Julie Urbanik draws directly from Wolch in her own work on Kansas City, Missouri. By tracing the ways that Kansas City has always been an assemblage of humans and nonhumans, Urbanik offers a rich case study upon which to articulate her idea of "cultural animal landscapes." Urbanik's innovative approach to historical animal geography highlights the promises and potentialities of new methodologies, exploring the narrative opportunities of digital media as a way to map the morphology of the cityspace and visually re-animate urban histories produced between humans and animals. Chris Philo and Ian MacLachlan's chapter also seeks to reveal that which has been hidden in animal geographies, but in a different register. They convincingly assert that slaughterhouses are missing from scholarly study in animal geography, avoided precisely because they traffic in death at a time when animal geography is invested in more-than-human liveliness. This chapter reminds historical animal geographers that the desire for dinner works to render unimaginable numbers of animals into blood, guts, and flesh; put differently, enmeshment often has darker ends. They invite animal geographers to historicize this practice, and in so doing, not forget that domination is very often the way that nonhumans experience human agency. Thomas Webb's chapter rounds out this section with his examination of the re-introduction of pigs in wartime London as a necessary source of food and waste disposal. Webb contends that this historical moment suspended the usual assertion that livestock belonged outside of the city, offering new spaces and ways for humans and nonhumans to relate to each other within urban environments. Here, too, porcine residents occupied creases within the urban environment, present and yet largely unseen to human residents.

Part III: The nation – historical animal bodies and human identities

Reorienting our view, Part III takes aim at the role that animals have had in co-producing discourses and materialities of the nation. Jennifer Mateer's chapter on elephants in India emphasizes their importance as actors and political subjects in military campaigns, as symbols of wealth, and as a means of transportation. In examining the different ways that elephants and humans have been entangled across time, Mateer offers a re-conceptualization of these animals as political partners in crafting human/nonhuman lifeworlds. Dominik Ohrem's work re-conceptualizes the American West in the eighteenth and nineteenth centuries. He describes the West less as an actual place and more as a site of transgression and experimentation, where one could work through the ontological ambiguity of what it means to be human and nonhuman. The last chapter in this section is written by Heidi Nast. Here, Nast considers the equivalencies made between coal miners and pit bulls in nineteenth-century Britain. Drawing on a two-volume Parliamentary report on colliery life from 1842, Nast traces the ways that coal miners were associated with animal work through a variety of labor practices, while seeking to suture the trauma of this treatment within dog-fighting pits.

Part IV: The global – imperial networks and the movements of animals

The final section in this book begins with another reprint from a luminary in human–animal studies: Harriet Ritvo. In "Migration, Assimilation, and Invasion in the Nineteenth Century," Ritvo explored the global movement of introduced animal species through colonization. Complicating articulations around the Columbian Exchange, Ritvo contends that even when animals were successfully transplanted from one place to the next (fraught projects characterized by uncertainty), they also often worked to reveal the fiction of human domination of the landscape and the unruly agency of all living things. David Lambert's contribution serves as an entreaty to historians of slavery to take the animal turn seriously. He invites the abandonment of what he names as the human exceptionalism of scholarship on slave societies and instead recognizes that both human and nonhuman labor made up colonial life. Finally, the book ends with an epilogue written by Stephanie Rutherford that, using the lens of the Anthropocene, offers a view into future directions for interdisciplinary research examining the confluence of geography, history, and animal studies.

When Faye Leerink, the geography editor from Routledge, invited us to submit an edited volume based on our sessions at the annual meeting of the AAG, we had no idea what kind of interest there might be in such a book. We were surprised and lucky, then, that we were found an innovative and inspiring community of scholars who were navigating the complex entanglements that form our more-than-human worlds over a broad expanse of space and time. Our meeting space has been infinitely enlarged and enriched by our involvement in this project.

References

Barua, M. (2016). Lively commodities and encounter value. *Environment and Planning D: Society and Space*, 34(4): 725–744.

Barua, M. (2017). Nonhuman labour, encounter value, spectacular accumulation: The geographies of a lively commodity. *Transactions of the Institute of British Geographers*, 42(2): 274–288.

Buller, H. (2014). Animal geographies I. *Progress in Human Geography*, 38(2): 308–318.

Buller, H. (2015). Animal geographies II: Methods. *Progress in Human Geography*, 39(3): 374–384.

Coleman, J. (2004). *Vicious: Wolves and men in America*. New Haven, CT: Yale University Press.

Collard, R.-C. (2012). Cougar-human entanglements and the biopolitical un/making of safe space. *Environment and Planning D: Society and Space*, 30(1): 23–42.

Collard, R.-C. (2014). Putting animals back together, taking commodities apart. *Annals of the Association of American Geographers*, 104(1): 151–165.

Datson, L. & Mitman, G. (eds.) (2005). *Thinking with animals: New perspectives on anthropomorphism*. New York, NY: Columbia University Press.

Fudge, E. (2017). What was it like to be a cow? History and animal studies. In E. Kalof (Ed.), *The Oxford handbook of animal studies* (pp. 258–278). New York, NY: Oxford.

Gillespie, K. & Collard, R.-C. (eds.) (2015). *Critical animal geographies: Politics, intersections and hierarchies in a multispecies world*. New York: Routledge.

Haraway, D. (2008). *When species meet*. Minneapolis: University of Minnesota Press.

Landry, D. (2008). *Nobel brutes: How eastern horses transformed English culture*. Baltimore, MD: Johns Hopkins University Press.

Lorimer, J. (2015). *Wildlife in the Anthropocene: Conservation after nature*. Minneapolis, MN: University of Minnesota Press.

Nance, S. (ed.) (2015). *The historical animal*. Syracuse, NY: Syracuse University Press.

Pearson, C. (2013). Dogs, history, and agency. *History and Theory: Studies in the Philosophy of History*, 52(4): 128–154.

Philo, C. & Wilbert, C. (eds.) (2000). *Animal spaces, beastly places: New geographies of human-animal relations*. London and New York: Routledge.

Ritvo, H. (2010). *Noble cows and hybrid zebras: Essays on animals and history*. Charlottesville, VA and London: University of Virginia Press.

Rothfels, N. (ed.) (2002). *Representing animals*. Bloomington, IN: Indiana University Press.

Rothfels, N. (2008). *Savages and beasts: The birth of the modern zoo*. Baltimore, MD: Johns Hopkins University Press.

Swart, S. (2010). *Riding high: Horses, humans and history in South Africa*. Johannesburg: Wits University Press.

Urbanik, J. (2012). *Placing animals: An introduction to the geography of human-animal relations*. Lanham, MD: Rowman & Littlefield.

van Dooren, T., Kirksey, E. & Munster, U. (2016). Multispecies studies. *Environmental Humanities*, 8(1): 1–23.

Wolch, J. & Emel, J. (1998). *Animal geographies: Place, politics, and identity in the nature-culture borderlands*. New York: Verso.

Woods, M. (2000). Fantastic Mr. Fox? Representing animals in the hunting debate. In C. Philo and C. Wilbert (Eds.), *Animal spaces, beastly places: New geographies of human-animal relations* (pp. 182–202). London and New York: Routledge.

Part I

The home – shared spaces of cohabitation

2   When did pets become animals?

Philip Howell

According to the American Society for the Prevention of Cruelty to Animals (ASPCA), there are currently around 80 million pet dogs in the U.S., 90 million or so cats, and well over 100 million fish, to take only the most common categories of companion animal (ASPCA, n.d.). The American Humane Association (AHA) estimates that over 60% of American households now own a pet, with dogs living in between 37% and 46% of households, and cats in the range of 30% and 39% (AHA, n.d.). The American Pet Products Association (APPA), which supplies the ASPCA figures, estimates the pet industry in the United States to be worth $63 billion, an expenditure that has roughly doubled in twenty years, taking inflation into account (APPA, n.d.). To many, these figures, country-specific as they are, speak to a growth in petkeeping closely associated with Western standards of affluence, and the cultural geographer Heidi Nast is hardly alone in considering what she calls the "recent emergence of pet-love" as a product of post-industrial society (Nast, 2006, p. 897). In a globalizing world, this culture of pets is increasing worldwide, particularly in the emerging economies. Brazil apparently now has the highest number of small dogs per capita, and India has the fastest-growing pet dog population; China, by size, is emerging as the largest pet-care sector (Euromonitor International, 2012; Cerini, 2016). Such an effusion of "pet-love" is unprecedented, and its significance in terms of economic geography and global ecology should not be underestimated. It tells us something equally important about ourselves, since petkeeping is recognized as "a barometer of the relationship between humans and the natural world" (Sykes, 2014, p. 139).

Much depends, however, on precisely what we mean by "pet," and these bald figures just as clearly betray the critical ambiguity of terms like "ownership" – and for that matter "household," or, by extension, "family." Statistics like these reproduce a specific conception of what a "pet" *is*: which is to say, a very specific, modern, and Western norm of animal companionship, of major significance but inevitably of only recent vintage. It is hardly surprising, if we define "pets" like this, that their history is so truncated. But the history and geography of petkeeping cannot be reduced to the recent growth and globalisation of "pet-love." If we follow the fashion in preferring the neutral and seemingly more objective term "companion" – rather than the loaded and problematic term "pet" – we straightforwardly have a much more inclusive story about humans and other animals.[1] This narrative reminds us that animals have been companions of humans in an enormous variety of ancient and ancestral cultures, and they can be found in contemporary hunter-gathering and simple horticultural societies, too (Serpell, 1987). We might be tempted to conclude that companion animals are entirely ubiquitous, their existence difficult or impossible to dissociate from the long historical geography of animal domestication, a process stretching back thousands or tens of thousands of years, serving as practical auxiliaries or because they satisfy emotional needs, or some combination of the two.

So the immediate problem for us as historical geographers is whether we endorse the view that pets are in some key sense "modern," and a creation of the contemporary West in particular, or else that companionship with animals is ancient and universal, so much a matter of "deep history" that it starts not to look like conventional history at all.[2] Though it will be seen that my own considered preference is for analysing the historical geography of this aspect of animal–human relationships in terms of the emergence of particular, distinctive regimes of petkeeping, especially as they reflect what the historian Susan Nance calls "animal modernity," these stark alternatives are as unsatisfactory as they are unavoidable (Nance, 2015a). The problem, as ever, is that we can hardly get around the fact that our historical relationship with companion animals encompasses their characterisation by human beings: "no people, no pets" is how Katherine C. Grier succinctly puts it (2006, p. 6). This does not mean that the definition of "pet" is at all easy. A generation ago, the historian Keith Thomas provided a helpfully sharp working definition: pets were animals that were allowed into the house; they were given individual names; and they were never eaten (Thomas, 1983). Subsequent scholarship has unsettled and unpicked this seeming good sense. How far we have come is clear from a recent attempt at a "lived definition" of pets, with all its qualifications and hesitancies:

> The most important quality pets share is that they have been singled out by human beings. Not all pets live indoors; large pets may not even live in the same location as their owners. Some pets do not, in fact, have names. And a few pets do eventually get eaten, which simply reflects the contingent status of the designation. Pets receive special attention intended to promote their well-being, at least as people understand that condition. Or we intend that they receive this attention.
>
> (Grier, 2006, p. 8)[3]

All the same, the first lesson for the historical geographer is that instead of imposing apparently objective terms like "companion animals" or "companion species" upon the subjects of our research, which simply begs the questions as to what kind of companionship might be involved, and whether it is reciprocated, and how we can access and describe it, we need to squarely face up to the process by which animals have become "pets."

When did animals become pets?

Let us begin then with what is a deceptively simple question: *when* did animals first become enrolled as companion animals or "pets"? One answer is that this is a matter not really for history but for prehistory. In a recent survey, for instance, Lisa Sarmicanic asserts that "Both the domestication of animals and the subsequent practice of pet-keeping have existed for thousands of years. From preagricultural times to the present, animals as helpmates and companions have played a vital role" (2007, p. 172). We could hardly dispute the longevity of these relationships, but the elision of domestication and petkeeping, and the imprecision of the word "subsequent" is worthy of emphasis. If we define petkeeping as "treating individual animals with indulgence and fondness," as do James Serpell and Elizabeth Paul in an influential contribution, it is plainly difficult to read this back into all forms of animal domestication, from the goat or the dog (the usual candidates for primacy) onwards (1994, p. 133). Rolling the historical geography of pets into that of domestication is likely to raise more questions than answers. It is extremely difficult to identify this pre-eminently "emotional" relationship into archaeological remains, artefacts, and other elements of material culture to which we have access, and of course the further back we go, the more cautious we have to be. "The object-traces of human involvement with pets, past and present, reveal the evolving contours of routine practices, the cultural assumptions that underlie them, and the complex, deep feelings that pet owners have about them," writes Katherine Grier (2014, p. 124), rightly, but it is hard to identify with clear confidence people's feelings for other animals, let alone a whole community's or culture's. It is not impossible, and we might reasonably examine the widespread evidence of animal burials as at least a preliminary guide to the emotional significance attached to favoured animals. Animals buried alongside people, in arrangements that indicate special bonds, can be put forward as particularly powerful evidence for the existence of "pets": the very earliest include that of a skeleton of a puppy found cradled in the hand of the human, buried in what is now Eynan, northern Israel, which dates from some 12,000 years ago, and, more recently and intriguingly, a fox buried with a female human found in Uyan-al-Hamman in northern Jordan, which was interred over 4,000 years earlier, the fox appearing to predate the dog, however fleetingly and abortively, as "man's best friend" (see, for example, Davis & Valla, 1978; University of Cambridge, 2011). But whilst zooarchaeologists and others are surely right to raise such coburials as suggestive of the presence of "pets" in a vast range of early societies, it is hard to be absolutely sure what such burials meant, and what these individual animals meant to their human companions in life, if indeed this focus on "companionship" is appropriate. As the archaeologist Naomi Sykes has recently cautioned, "Even if we were able to classify confidently a set of animal remains as having derived from a 'pet,' this label does not adequately account for the likely complexity of the human–animal engagement" (Sykes, 2014, p. 133). Similarly, whilst the psychologist Hal Herzog has the confidence to assert that "Archaeological evidence of pet-keeping goes back

12,000 to 14,000 years for dogs and perhaps 9,000 years for cats," he immediately qualifies this by saying, in his own breezy fashion, that

> If we were to discover, say, the 25,000-year-old fossil remains of a man cradling a baby monkey, we would not be able to tell if the animal was the dead guy's pet or if the monkey was placed in his grave as a snack in the afterlife.
>
> (Herzog, 2011, p. 88)

We might extend this caution to the tricky question of whether petkeeping is natural or innate, and what we might mean by this. To turn again to Herzog's *Some We Love, Some We Hate, Some We Eat*, the author categorically insists that "The human being is the only animal that keeps members of other species for extended periods of time purely for enjoyment," noting the evidence for other animals keeping "pets" only to dismiss it as a phenomenon of captive animals – or at least that any evidence to the contrary is simply evidence that proves the rule (ibid, p. 87). Herzog elsewhere restates his position thus:

> I suspect that human-style pet love requires a combination [of] anthropomorphism and learned cultural values found in only one species – ours. Anthropomorphism enables us to empathize with non-human creatures, and cultural values give us permission to fall in love with some types of animals.
>
> (Herzog, 2012)

The point here is not to pretend to an expertise and authority that I do not possess, but merely that we have moved in such anthrozoological science quickly from confident assertion to at best informed guesswork, in the manner recently critiqued by Vinciane Despret and Bruno Latour as "academocentrism" (whereby a properly disciplined animal science prides itself on simply eliminating alternative accounts) (Despret, 2016). Here, petkeeping is said to require anthropomorphism, by definition an exclusively human practice, *ergo* only human beings keep pets. The natural history of petkeeping, if this is how we should put it, is an open question, and we should not suggest otherwise by such patently anthropocentric sleights of hand.

None of this is meant as an argument for ignoring the evidence of the antiquity of animal companionship, and there is enormous potential in fields such as zooarchaeology and anthrozoology to shed light on the history and geography of petkeeping. It is indeed quite safe to say that companion animals of some sort or other can be found in almost any historical or prehistoric culture that we care to consider, with evidence in many cases that these animal companions were treated in the ways that we now associate with the word "pet." Hardly anyone is more expert in this regard than James Serpell, and he has argued that fondness for pets exists in all societies (Serpell, 1989, p. 13). There is little reason not to think that people have loved and grieved for companion animals in a manner we find perfectly familiar. Many ranks and classes of Greeks and Romans kept pets, for example, with Michael MacKinnon judging that emotional attachments to at least some of these animals ran deep.[4] The evidence for the European Middle Ages suggests much the same, with very rich textual and visual sources to draw upon – even if many of these sources are critical of the practice of petkeeping, particularly where the pet guardians were monks and nuns.[5] By the time we come to the edge of the modern world, as we conventionally recognise it, sentiment and emotion seem utterly incontrovertible. To give just one example: Isabella d'Este, the "first lady of the Renaissance," staged a funeral in 1510 for her cat Martino, who was buried in a lead casket in his own tomb and was quickly followed to the afterlife by one of his mourners, Isabella's beloved dog Aura. As her secretary reported,

> It is not possible to speak of Madama's grief; there is so much of it. Anyone who knows the love she bore the dog can well imagine it. And much was deserved as Aura was the prettiest and most agreeable little dog that ever there was. Her ladyship was seen crying that evening at dinner, and she couldn't talk about it without sighing. Isabella cried as if her mother had died and it was not possible to console her.
>
> (cited in Walker-Meikle, 2012, pp. 33–34)

Such sentiments are what we have come to expect of petkeeping, and of what in the drier academic parlance is labelled "zoophilia." As the zooarchaeologist Nerissa Russell puts it, "Pet keepers simply like having animals around" (2012, p. 263).

We must not erase, however, the specific significance of petkeeping in this focus on a basic and universal human affection for animals, however engaging these examples. James Serpell's and Elizabeth Paul's (1994, p. 129) much-cited definition of "pet" – "animals that are kept primarily for social or emotional reasons rather than for economic purposes" – is difficult to endorse across the ages and across all cultures, since the distinctions it makes or implies, between the secular and spiritual, between the material and the sentimental, between violence and love, and so on, are inherently ambiguous, even today. We should be wary too of suggesting that affection for individual animals, or perhaps only individual animals in their immaturity, is the same as enduring affection for "companion species" – let alone "animals" in a collective and abstract sense.[6] We are constantly reminded that we need to be able to understand why in some times and places even our most familiar pets, such as cats and dogs, were or were not called into service. All of these species have different agencies and capacities to perform in the role or roles of "companion," qualities that may be brought out and valued in one culture in one period but not in another. We might advise then that the historical geographies of, say, companion cats and dogs, need to be considered separately, and so too the various birds, reptiles, amphibians, fish, exotics and even insects that have attracted attention at different times as "pets" – even before we think of turning to the varieties, breeds and other classes of subspecies. In short, examples of affection for individual animals do not necessarily imply the kind of trans-species relationships that we are thinking of when we conceive of "pets." Some writers have suggested that we use the term "personal animals" as an alternative, precisely to avoid ahistorical confusion.[7]

Above all, we need to draw a distinction between the very widespread practice of keeping pets and its cultural authorisation or legitimacy. Here we have the most obvious differences between ancient and modern petkeeping. Ingvild Gilhus states quite bluntly, for example, that "Graeco-Roman society was not a pet-keeping society" (Gilhus, 2006, p. 29). Naomi Sykes similarly argues that "for much of the medieval and post-medieval periods, pet-keeping was not something to shout about" (2014, p. 139). The question we started with thus has to be set aside, in order to consider the nature and the social acceptance of the relationship between the pet and its guardian. We need to look to long-term shifts in attitudes toward animals, at relationships between people, and at the regimes in which interspecies relationships are alternatively approved or denigrated.

When did pets become animals?

We can come at these issues by simply reversing the terms, and asking "when did pets become animals?" I realize that this suggests the worst kind of academic sophistry, but it is a useful move because it reminds us that the concept of the "pet" refers not to a thing or an object, but rather to a relationship. The cultural geographer Yi-Fu Tuan (1984), in a vital and pioneering foray, still of immense relevance, taught us that what we call petkeeping is essentially a relation of *power* – for him, the power of humans over their environments, broadly conceived, taking in nonhuman animals for sure, and also nonhuman nature in its entirety, but, most pertinently, extending also to human beings. In this regard, the point is not to lamely universalise but to locate, to show how pets were, in Tuan's sense, "made" – that is, to trace the circumstances in which gardens and landscapes, or servants, women, children – or nonhuman animals – were drawn up into the ethos of petkeeping and its distinctive dyad of "dominance" and "affection." Now, for all his brilliance, Tuan is not much of a guide for the historical geographer of nonhuman animals: "In Europe, society as a whole seemed to show a warmer feeling toward domestic animals from the seventeenth century onwards" is just about as specific as he gets, whilst his shrug of explanation – the increasing distance of people from nature in urban and industrialising societies – is as inadequate as it is conventional (ibid, p. 111). But if this account is historically thin, Tuan's intuitions are still instructive: that is, the fact that "pets" could be all sorts of things, including people, and only in the modern age, our age, has the nonhuman animal become the pet's more or less exclusive icon.

There seems to be every reason to look to the early modern period for the beginnings of this process of simplification and specificity. If the etymology of English is used as a guide, for instance, we can locate the pet reasonably precisely: the word means an animal kept for pleasure or companionship, and seems to have been used not much earlier than the sixteenth century, possibly entering the language as late as the early eighteenth century (Oxford English Dictionary, 2017). Moreover, this sense sat alongside, for a long time, the fact that "pet" could refer to a spoiled or indulged child, or any favourite. Even when only nonhuman animals are invoked, the difference between an individual *domesticated* animal, such as a lamb reared by hand, and the *domestic* pets with which we are more familiar, is striking. It would be unwise to generalise too quickly from the world of words, and I am aware of being Anglocentric as well as Eurocentric, but this semantic history reinforces the impression of a later transformation, the relative novelty of the animal companion as "pet," even its "invention." And whilst we know that there were companion animals long before, the early modern period does seem to herald a quickening of petkeeping, with the rapid objectification of domestic livestock in the pre-industrial period matched by a process in which pets were "increasingly endowed with quasi-human status" (Raber, 2007, p. 73). By the advent of the modern world as we usually understand it, we find not only a greater host of animal "pets," but much evidence for the indulgence of these animals: "By 1700 all the symptoms of obsessive pet-keeping were in evidence," says Keith Thomas, thinking of England, adding that these animals often seem to be treated better than the servants (1983, p. 117). The historian Ingrid Tague (2015) agrees, and has made a recent bid for identifying the invention or making of the pet in eighteenth-century England. The terms of Tague's argument bear emphasis, in contrast to Thomas's, for it is not merely the keeping of animal companions, nor even the eccentricities of petkeeping, that are at issue, but rather the developing societal acceptance of pets. There is a strong case to be made that near the beginning of Tague's period, pets were preeminently portrayed as useless luxuries, the playthings of a corrupted aristocracy and their womenfolk in particular. As in the middle ages, if we recall Naomi Sykes's characterisation, petkeeping was hardly something to shout about – and indeed pet guardians found themselves much shouted at. Tague documents the angry concomitant of a rise in the ownership of pets, which is the vicious condemnation of these animals and their owners, often taking the form of the most blatant misogyny. Such quintessentially ladies' pets as lapdogs were scorned as frivolities, fashion accessories, and wasteful indulgences. Worse, they could be pilloried as a perversion of proper human relations, for a married woman's natural affections were supposed to be directed to her husband and children, not her lapdog. As Alexander Pope puts it, in *The Rape of the Lock* (1712, Canto III): "Not louder Shrieks to pitying Heav'n are cast,/When Husbands or when Lap-dogs breathe their last." If the pet entered history at the advent of modernity, then, it was principally as a kind of monstrosity. This finger-wagging admonitory mode has never gone away, but it is hard to miss the apogee of discontentment and derision directed at pets and pet guardians at the advent of modernity.

How then did pets become respectable – or at least more so? Tague claims to identify the change in her period, using for instance the fact that owners increasingly desired to be portrayed with their pets – pet dogs in particular – something she sees as evidence of individuation and affection. By the time we arrive at the age of Victoria, who is famously pictured with her family, with their pet dogs frolicking alongside their children, in the very image of a proper household, we are surely a full stride closer to the contemporary idealisation of the "family pet."[8] It is impossible to do full justice to this change in attitudes, but we can certainly link this to the growth of a culture of sentiment and sensibility, to Romanticism, to the rise of an anti-cruelty movement and what we can recognise as pro-animal ideas and practices; a culture of petkeeping was surely nurtured within such epochal shifts.[9] In the place of a perverted intruder, a kind of witch's "familiar," the pet increasingly comes to serve as a subordinate but valued member of the family: an icon of domesticity, valued for teaching children the duties and responsibilities of care, and a living zoology lesson and a reminder of the idealised natural world into the bargain (see, for example, Mason, 2005).

As I have been at pains to note, however, emotional attachments to individual pets can hardly be used to chart these wider changes, even if the sheer weight of numbers feels telling. Simply put, if we define petkeeping principally as an *emotional* bond, it is hard to be at all precise about chronology, certainly not with anecdotes of affection. Love for pets is certainly evident in the eighteenth century, as Tague shows – why should petkeeping be satirised for its emotional wastefulness unless this love existed in the first place? The early eighteenth-century correspondence of Lady Isabella Wentworth, to take a well-known example, fulsomely demonstrates the depth of affection that many owners felt toward their pets, which in Isabella's case included Fubs (a dog), Pug (confusingly, a monkey), and Puss (no real surprise here, a cat).[10] On Fubs's death in 1708, Isabella lamented: "Sure of all of its kind there never was such a one nor never can be, so many good qualities, so much sense and good nature and cleanly and not one fault; but few human creatures had more sense" (cited in Tague, 2015, p. 206).[11] Ingrid Tague and others draw deeply on such examples, quite rightly. But the fact that this language can be used in similar circumstances, but to quite different effect, is obvious from the fact that exactly a century later, Lord Byron eulogised his dead dog Boatswain in superficially very similar language, as "one/who possessed Beauty without Vanity/ Strength without Insolence/Courage without Ferocity/and all the virtues of man without his vices."[12] But Byron's famous lines are not all that they seem: they were probably provided by a friend, and the aim in his much-cited epitaph is almost certainly satirical, misanthropic rather than straightforwardly sentimental, perhaps in uncomfortably large part mockery. The point is that we should not leap to see either Isabella Wentworth, or Byron, as somehow representative of the spirit of the age, part of the annunciation of "pet-love."

We need to look elsewhere to identify the historical and geographical specificity of *modern* petkeeping. I am hardly unbiased, but there is a strong case to be made for a focus on the nineteenth rather than the eighteenth century – a case based less on sentimentalism than on the evidence for an institutionalisation of the pet (Howell, 2015). In the second half of the nineteenth century, if we train our glass on dogs, there are a whole series of milestones on the remarkable journey to the modern pet. We have in England the first true dog shows, at Newcastle in 1859, or more plausibly Birmingham in 1860 (Howell, 2016). The world of pedigree dog breeding was speedily formalised, with perhaps "the first modern dog" (in terms of conformation to the new breed standards) appearing in 1865, and the Kennel Club was established a few years later to regulate the business of breeding and showing (University of Manchester, 2013). Thinking in terms of the pet industry, in or around 1860 the world's first dedicated mass-produced dog food, Spratt's famous "X Patent" dog biscuits, was launched (one of Spratt's employees, a

Mr. Charles Cruft, was to lend his name in 1891 to the most famous dog show in the world). All of these events suggest the rise of an infrastructure for the modern pet and a newly institutionalised type of petkeeping. On these grounds, it is I think acceptable hyperbole to say that the Victorian age was when the modern dog was "invented," and perhaps by extension – though this is more problematic given my advice to proceed species by species – the modern pet (Homans, 2012, p. 1).

I want to make one final claim in this regard, however. The year 1860 also saw the founding of the Battersea Dogs' Home, and this institution testifies to the desire to place the pet firmly in the security of the home. This pioneering animal rescue society "restored" lost dogs to their owners and "rehomed" as many other "home-less" dogs, including street and stray dogs, as it could. The ethos of these shelters and refuges – and many others followed Battersea's lead – demonstrates how central domesticity was to the destiny of the companion animal or pet, by the time we reach the nineteenth century. If dogs are at all representative, the institution of petkeeping follows the injunctions of a cultural geography that placed "pets" firmly in the home, with the concomitant that dogs were "out of place" in the public streets. The consequences for the unwanted stray dog, and those thousands of dogs who were not deemed capable of serving as pets, was dire. Street dogs' lives were increasingly inauspicious: being without a home was the same as being without ownership and without all the protections that being *property* conferred. The stray dog was ever more vulnerable to being policed out of the public streets and out of this world altogether – many dogs' homes today are forced to be, in the uncomfortable and oxymoronic modern parlance, "kill shelters." This aspect of petkeeping we neglect if we focus on "pet-love" alone. The "pet" was promised affection, albeit conjoined with dominance, but his or her alter ego, the "stray," was greeted with dominance in the form of destruction. Our world of pets and petkeeping has many habitual cruelties, as we know, but the fate of unwanted animals (we might think of them as "failed pets") is not incidental: out of the four million dogs and over three million cats that now enter shelters in the U.S. every year, one out of every three dogs and two out of every five cats have their lives ended there (ASPCA, n.d.). The flipside of "pet-love," what Andrei Markovits and Katherine Crosby (2014) call the "discourse of compassion," is the cold reality of animal killing. In this regard, the question, "When did pets become animals?" is probably impossible to separate from the wider process by which all animals became killable on an industrial scale.

Conclusions

Companion animals have always existed, then, but such companions only became "pets" in the modern era. To speak of the historical geography of pets is not a mat-ter of amassing or mapping human beings' emotional attachments toward other animals. It is rather to chart the development of a society, or culture, or regime, in which the relations between humans and companion animals – as individuals and as species – is normalised and made more or less respectable. Above all, we have to consider the specific conditions under which this "pet-love" is allowed and authorized. Only when such iconic animals as dogs and cats are reduced to the status of property, with all that this portends, can we speak confidently of the age of the pet.

Notes

1 See Podberscek, Paul, and Serpell (2000). On the alternative, "companion species," see Haraway (2003).
2 For this kind of work on animal history, see, for instance, Braje (2011). Those who prefer conventional histories of animals should look first to Kalof and Resl (2007). Reflections on the history of animals can be found in the following collections: Brantz (2010) and Nance (2015).
3 For a further attempt at definition that students may find useful, see Grier (2014).
4 See, for instance, Bodson (2000), Gilhus (2006), Kalof (2007), and MacKinnon (2014), 279.
5 See, for instance, Thomas (2005) and Walker-Meikle, K. (2012).
6 I draw here on Jacques Derrida's influential critique, in Derrida (2008).
7 This suggestion comes from Gilhus (2006), 29.
8 For the link to the family, see Shell (1986); geographers should also consult Power (2008).
9 For *some* historical work that considers these complex questions, see Boddice (2009), Kenyon-Jones (2001), and Preece (2005).
10 For Wentworth, see Tague (2015), 200–209.
11 Letter, November 16, 1708. This correspondence can be consulted in the British Library.
12 The epitaph is inscribed on a memorial in the grounds of Newstead Abbey, Nottingham-shire, UK; the memorial and its meaning is particularly well discussed in Chapter 1 of Kenyon-Jones (2001).

References

American Humane Association (no date). U.S. pet (dog and cat) population fact sheet. Retrieved February 7, 2017, from www.bradfordlicensing.com/documents/pets-fact-sheet.pdf.
American Pet Products Association (no date). Pet industry market size and ownership statistics. Retrieved February 7, 2017, from www.americanpetproducts.org/press_industrytrends.asp.
American Society for the Prevention of Cruelty to Animals (no date). Pet statistics. Retrieved February 7, 2017, from www.aspca.org/animal-homelessness/shelter-intake-and-surrender/pet-statistics.
Boddice, R. (2009). *A history of attitudes and behaviours toward animals in eighteenth- and nineteenth-century Britain: Anthropocentrism and the emergence of animals.* Lewiston, NY: Mellen.
Bodson, L. (2000). Motivations for pet-keeping in ancient Greece and Rome: A preliminary survey. In A. L. Podberscek, E. S. Paul, and J. A. Serpell (eds.), *Companion animals and us: Exploring the relationships between people and pets* (27–41). Cambridge: Cambridge University Press.
Braje, T. J. (2011). The human-animal experience in deep historical perspective. In C. Blazina, G. Boyra, and D. S. Shen-Miller (eds.), *The psychology of the human-animal bond* (63–80). New York, NY: Springer.
Brantz, D. (ed.) (2010). *Beastly natures: Humans, animals, and the study of history.* Charlottesville, VA: University of Virginia Press.

Cerini, M. (2016, March 23). China's economy is slowing, but their pet economy is boom-ing. Retrieved February 7 2017, from www.forbes.com/sites/mariannacerini/2016/03/23/chinas-economy-is-slowing-but-their-pet-economy-is-booming/#383aaaf37a48.

Davis, S.J.M. and Valla, F. R. (1978). Evidence for domestication of the dog 12,000 years ago in the Natufian of Israel. *Nature* 206: 608–610.

Derrida, J. (2008). *The animal that therefore I am.* New York, NY: Fordham University Press.

Despret, V. (2016). *What would animals say if we asked the right questions?* Minneapolis, MN: University of Minnesota Press.

Euromonitor International (2012, November 13). The dog economy is global – but what is the world's true canine capital? Retrieved February 7, 2017, from www.theatlantic.com/business/archive/2012/11/the-dog-economy-is-global-but-what-is-the-worlds-true-canine-capital/265155/.

Gilhus, I. S. (2006). *Animals, gods and humans: Changing attitudes to animals in Greek, Roman and early Christian ideas.* London: Routledge.

Grier, K. C. (2006). *Pets in America: A history.* Chapel Hill, NC: University of North Carolina Press.

Grier, K. C. (2014). The material culture of pet keeping. In G. Marvin and S. McHugh (eds.), *Routledge handbook of human-animal studies* (124–138). London: Routledge.

Haraway, D. (2003). *The companion species manifesto: Dogs, people, and significant otherness.* Chicago, IL: University of Chicago Press.

Herzog, H. (2011). *Some we love, some we hate, some we eat: Why it's so hard to think straight about animals.* New York: HarperCollins.

Herzog, H. (2012, July 25). Are humans the only animal that keep pets? Retrieved February 7, 2017, from www.huffingtonpost.com/hal-herzog/are-humans-the-only-anima_b_1692004.html.

Homans, J. (2012). *What's a dog for? The surprising history, science, philosophy and politics of man's best friend.* London: Penguin.

Howell, P. (2015). *At home and astray: The domestic dog in Victorian Britain.* Charlottesville, VA: University of Virginia Press.

Howell, P. (2016). June 1859/December 1860: The dog show and the Dogs' Home. In D. N. Felluga (ed.), *BRANCH: Britain, representation and nineteenth-century history.* Extension of Romanticism and Victorianism on the Net. Retrieved February 7, 2017, from www.branchcollective.org/?ps_articles=philip-howell-june-1859december-1860-the-dog-show-and-the-dogs-home.

Kalof, L. (ed.) (2007). *A cultural history of animals in antiquity.* Oxford: Berg.

Kalof, L. and Brigitte Resl, B. (eds.) (2007). *A cultural history of animals,* 6 volumes. Oxford: Berg.

Kenyon-Jones, C. (2001). *Kindred brutes: Animals in Romantic-period writing.* Aldershot: Ashgate.

MacKinnon, M. (2014). Pets. In G. L. Campbell (ed.), *The Oxford handbook of animals in classical thought and life* (269–281). Oxford: Oxford University Press.

Markovits, A. S. and Crosby, K. N. (2014). *From property to family: American dog rescue and the discourse of compassion.* Ann Arbor, MI: University of Michigan Press.

Mason, J. (2005). *Civilized creatures: Urban animals, sentimental culture, and American literature, 1850–1900.* Baltimore: Johns Hopkins University Press.

Nance, S. (2015a). *Animal modernity: Jumbo the elephant and the human dilemma.* Houndmills: Palgrave Macmillan.

Nance, S. (ed.) (2015b). *The historical animal.* Syracuse, NY: Syracuse University Press.

Nast, H. J. (2006). Critical pet studies? *Antipode* 38(5): 894–906.

Podberscek, A. L., Paul, E. S., and Serpell, J. A. (eds.) (2000). *Companion animals and us: Exploring the relationships between people and pets* (27–41). Cambridge: Cambridge University Press.

Power, E. (2008). Furry families: Making a human-dog family through home. *Social & Cultural Geography* 9(5): 535–555.

Preece, R. (2005). *Brute souls, happy beasts and evolution: The historical status of animals*. Vancouver, BC: University of British Columbia Press.

Raber, K. (2007). From sheep to meat, from pets to people: Animal domestication 1600–1800. In M. Senior (ed.), *A cultural history of animals in the age of Enlightenment* (73–99). Oxford: Berg.

Russell, N. (2012). *Social zooarchaeology: Humans and animals in prehistory*. Cambridge: Cambridge University Press.

Sarmicanic, L. (2007). Bonding: Companion animals. In M. Bekoff (ed.), *Encyclopedia of human-animal relationships* (162–174). Westport, CT: Greenwood Press.

Serpell, J. A. (1987). Pet-keeping in non-western societies: Some popular misconceptions. *Anthrozoos* 1(3): 167–174.

Serpell, J. A. (1989). Pet-keeping and animal domestication: A reappraisal. In J. Clutton-Brock (ed.), *The walking larder: Patterns of domestication, pastoralism, and predation* (10–21). London: Unwin Hyman.

Serpell, J. A. and Paul, E. (1994). Pets and the emergence of positive attitudes to animals. In A. Manning and J. Serpell (eds.), *Animals and human society: Changing perspectives* (127–144). London: Routledge.

Shell, M. (1986). The family pet. *Representations* 15(Summer): 121–153.

Sykes, N. (2014). *Beastly questions: Animal answers to archaeological issues*. London: Bloomsbury.

Tague, I. H. (2015). *Animal companions: Pets and social change in eighteenth-century Britain*. New Haven, CT: Yale University Press.

Thomas, K. (1983). *Man and the natural world: Changing attitudes in England, 1500–1800*. London: Allen Lane.

Thomas, R. (2005). Perception versus reality: Changing attitudes towards pets in medieval and post-medieval England. In A. Pluskowski (ed.), *Just skin and bones? New perspectives on human-animal relationships in the historical past* (93–101). Oxford: Archaopress.

Tuan, Y.-F. (1984). *Dominance and affection: The making of pets*. New Haven, CT: Yale University Press.

University of Cambridge (2011, January 31). Was the fox prehistoric man's best friend? Retrieved February 7, 2017, from www.cam.ac.uk/research/news/was-the-fox-prehistoric-mans-best-friend.

University of Manchester (2013, March 6). First modern dog discovered. Retrieved February 7, 2017, from www.manchester.ac.uk/discover/news/article/?id=9636.

Walker-Meikle, K. (2012). *Medieval pets*. Cambridge: Boydell.

3 The entwined socioecological histories of the Sawtelle, California war veterans and the animal "menagerie" at the Pacific Branch Soldier's Home (1888–1918)

Teresa Lloro-Bidart

The United States Congress established the National Home for Disabled Volunteer Soldiers (NHDVS) in 1865 as a network of federally funded domiciles and hospitals designed to house and care for disabled Union veterans, though by 1884 these restrictions were lifted and any aged veterans were able to qualify as disabled and receive services (Kelly, 1997; Logue, 1992; Wilkinson, 2013b). These homes cropped up all over the nation, with California opening a west coast branch, referred to as the "Pacific Branch," in the Sawtelle district of west-side Los Angeles in May 1888, an area that flanks the city of Santa Monica. By the end of the Pacific Branch Home's first fiscal year the following June, over 300 men resided at the facility (Wilkinson, 2013b), and by the end of 1911, the *Los Angeles Times* reported that 4,000 veterans were using services provided there (Unknown Author, 1911). The surrounding Santa Monica community quickly noted the veterans' presence, as their dark blue uniforms marked their "invalidism and institutionalism" and also suggested they were poor, not only fiscally, but also in character and morals (Wilkinson, 2013b, p. 199).

As Wilkinson (2013a, 2013b) illustrates, little has been written about members of the Home or the city of Sawtelle's veteran residents, which reflects a general trend in the historiography regarding Union veterans living in the West. Furthermore, the academic literature has been completely silent about human–nature relationships at the Home, despite the fact that contemporary popular sources describe a variety of more-than-human species who also resided there (see, for example, Driggs, 2011; van de Hoek, 2005). This chapter, therefore, investigates human–animal relationships at the Home using Hanson's middle landscape concept, which theorizes "gardens, parks, or other natural landscapes" as "places that integrate nature and culture, joining pastoral scenery and civilization" with "moral landscape[s]" (2002, p. 17). While having historical significance that pre-dates the Victorian era, Hanson demonstrates how middle landscapes became particularly significant in the late nineteenth century, contemporary to the rise of public parks and zoological gardens as they became places to "uplift" the working class through exposure to middle-class values (2002, p. 20). Although Hanson (2002) primarily discusses traditional zoos in her text – that is, those that were expansive, purposeful, public, and meant to serve as refuges from the city – I expand her work here to show that some of these goals were translated to the Home. Besides citygoers' efforts to curtail the veterans' behaviors through outlawing saloons within the city's boundaries and gerrymandering school voting districts, for example, the Home itself appears to have used nature, including animals, as a less overt tactic to civilize the men to contemporary middle-class values.[1]

Evidence suggests that the Home provided many such opportunities for accessing nature, including maintenance of the Home's farm, garden, and even "animal menagerie" or "zoo." Although several different archival sources chronicle the experiences of animals at the menagerie and their roles in shaping veteran life at the Home, these materials are decidedly limiting, not only because they are few in number, but also because they tell stories about the animals from a specific and narrow human perspective that does not necessarily "speak on behalf of the animal" (Woods, 2000, p. 199) or adequately capture animal agencies (Hribal, 2007). Thus, constructing animal histories "from below," or what Hribal refers to as histories that explore how animals co-construct their own lives as "laborers, prisoners, or resistors" (2007, p. 102), for example, is in many ways quite fraught. Yet, the vivid descriptions of animals at the Home, especially in local newspapers, suggest that they participated in complex ways in civilizing the veterans to the middle-class life of the Santa Monica community, sometimes because of their captivity (e.g., throwing rocks and biting) and at other times in spite of it (e.g., singing). Although clearly not written from the animals' viewpoints, this study's archival materials do vibrantly capture animal action, including how the Home worked to limit that action through the purchase of materials like cages and netting, the former of which one specific monkey, Joe, appears to have resisted. If viewed as performative aspects of their local (and oppressed) geographies, rather than as innate qualities (Hovorka, 2015), and if understood with one of Fudge's central arguments in mind – that an important aspect of "reading animals" is "through the ideas, attitudes, worries of the human agents" (p. 106) – such behaviors can be interpreted relationally with human action and as imbued with specific meanings that facilitate the writing of animal history, even when little documentation of animal life remains.

Thus, after introducing the middle landscape concept and providing a brief history of the origins of the Pacific Branch, I discuss how the Home appears to have leveraged a public park-like atmosphere and the "animal menagerie" or zoo to "civilize" and uplift the veterans and instill in them good health and morals through direct contact with animals. In this specific context, hospital administrators appear to have evicted the uncivilizable and unruly monkeys at the Home, who did not represent the middle-class ethos of the surrounding cityscape, which was fraught with many of the same anxieties characteristic of urbanizing spaces throughout the United States. The Home's bird collection, which provided "sweeter music" and was perceived as offering tranquility to the area's residents who were most poor in character, morals, and habits – the veterans – lasted much longer. As a result, the ideals and material reality of the middle landscape were not embedded within the city but within the boundaries of the Pacific Branch Home.

The "middle landscape"

Zoos serve as institutions that reflect how various human cultures view nature and their place within it (Anderson, 1995; Braverman, 2013; Hanson, 2002; Rutherford, 2011). In her text *Animal Attractions*, Hanson argues that the "new zoos," which began to spread across the country in the 19th and early 20th centuries – contemporary to the establishment of the NHDVS – "quickly became emblems of civic pride, an amenity of every growing and forward-thinking municipality analogous to other institutions such as art museums, natural history museums, and botanical gardens" (2002, p. 3). Furthermore, while zoos initially emphasized entertainment and have since embraced a more explicit focus on education, zoos have always already been "educational" insofar as they reflect the sociocultural and socioecological values of the societies in which they are placed and therefore implicitly or explicitly teach people who interact with the captive animals residing there. As Hanson notes, in the late nineteenth and early twentieth centuries, zoos emphasized a "middle-class ethos," and together with the city parks within which they were often constructed, constituted "middle landscapes." These spaces purposefully joined "pastoral scenery and civilization" (2002, p. 17) in an effort to provide refuge for the middle class, whose values were reflected in these landscapes, and also to "uplift the working class" with the "good health and morals" one could find through contact with and appreciation of nature (2002, p. 20).[2]

Although the expansiveness of the Pacific Branch's collection of animals is not known, it was referred to as a "zoo" or "menagerie" in several local newspaper articles (Unknown Author A, 1896; Unknown Author B, 1902) and in a small historical booklet detailing life at the Home (Soldiers' Home Booklet, 1900). The zoo apparently contained a diverse array of mammal and bird species that the Home invested federal dollars in maintaining, at least until 1913 (Minutes of the Council of Administration Pertaining to Post Funds, 1901–1913). Indeed, veterans were sometimes referred to as "keepers" and the animals as "specimens" (Unknown Author A, 1903) or as part of a "collection," "menagerie," or "animal garden" (Minutes of the Council of Administration Pertaining to Post Funds, 1901–1913; Unknown Author A, 1896; Unknown Author B, 1896; Unknown Author B, 1901), suggesting a zoo or museum-like atmosphere (Hanson, 2002; Rader & Cain, 2014). The Home also logged multiple purchases for wire netting, canvas hoods, and cages (Minutes of the Council of Administration Pertaining to Post Funds, 1901–1913), even though bits of other evidence align more with the animals as pets (Unknown Author, 1891), which Grier (2006) defines as domesticated animals one would purposefully keep for companionship. Since the boundaries between zoo animal and pet have not historically been neatly defined (e.g., Fudge, 2008; Hanson, 2002), and given that the Pacific Branch Home complexly blurred private and public life, it is likely that some of the veterans had an equally ambiguous and complex relationship with the Home's animals. Hanson (2002) notes, for example, that during the early twentieth century, zoos would often temporarily board pets for their owners, who viewed these stays as "privileged summer camps" (p. 55); in many instances, zoos also served as dumping grounds for unwanted pets. While evidence suggests that some of the veterans provided actual care for the animals on a local scale, the Pacific Branch Home (and by extension the federal government) fiscally provided for both, and ultimately the administrators and managers got to decide whether the animals (and the veterans) got to stay (Sawtelle Disabled Veterans Home, Los Angeles Case Files, 1888–1933; Unknown Author, 1901a; Unknown Author C, 1901, Unknown Author, 1908).

In the following analysis, therefore, I use Hanson's (2002) "middle landscape" to theorize the Pacific Branch Home's animal collection as a tactic to "civilize" and "uplift" the veterans and instill in them good health and morals through direct contact with nature. I demonstrate that it was not contact with any sort of nature that was thought to provide this uplift, but specifically the Home's bird collection, which appeared to be easier to control and maintain than the Home's monkeys and even the Home's farm animals. Households during the nineteenth and twentieth centuries often kept caged birds, who were viewed as "natural models for [the] middle-class family life" (Grier, 2006, p. 46) that was emblematic of the Santa Monica community the Home sought to socialize the veterans into after their service. In this case, instead of a public space (e.g., zoo) serving to inculcate moral values that would ideally translate to the domain of the private sphere (e.g., the home), the Pacific Branch's encompassing concept of "Home" worked to provide residents with certain kinds of nature perceived as appropriate, entertaining, and civilizing. That is, the middle landscape was embedded within the Home.

The Pacific Branch "Home"

Although there were many proposals to serve as the site of the Pacific Branch Home throughout the state, since it was viewed by various area elites as a vehicle to "secure the economic future for their city that the construction, operation, and constant supply of goods, services and labor necessary to build and operate an NHDVS branch would entail," the Board of Managers ultimately chose Sawtelle at the end of 1887, due to an enticing offer from the millionaire Senator from Nevada, John Percival Jones, and his business partner, Colonel Robert Symington Baker (Wilkinson, 2013b, p. 193).[3] Together, the "Baker and Jones offer" included 300 acres of land, a water supply, and $100,000 cash – Jones, in particular, was eager to develop the site in order to support his other local business ventures, which included turning the Santa Monica region into a bustling and prosperous commercial area (Wilkinson, 2013a). Area elites, therefore, viewed the veterans as a potential revenue stream that would fulfill their visions of further urbanizing Santa Monica.

As Wilkinson highlights, although the facility was called a "home," it "bore little resemblance to the Victorian ideals of what constituted one" since residents initially wore uniforms, lived in barracks, ate communally, and were prohibited from leaving for long periods of time (2013a, p. 192). While this high level of strict discipline eventually waned and veterans became more fully integrated into the city, the Home nevertheless continued to work to regulate the behaviors of veterans in multifaceted ways. Indeed, Kelly posits that the entire National

Home system was effectively an asylum that provided "modern, comfortable, and humane" care (1997, p. 7) and notes that "residents abandoning a branch were classified as deserters" (1997, p. 141). Soldier case files, for example, revealed multiple instances of deserters requesting re-admission to the Home so that they could resume access to provided services (Sawtelle Disabled Veterans Home, Los Angeles Case Files, 1888–1933). Although Kelly highlights that martial citizenship, which is the notion that citizenship within a society can be gained through military service and sustained through the welfare state "insinuated itself into the culture and economy of a number of American communities" (1997, p. 2), the Sawtelle case in some ways complicates this assertion. The local Santa Monica community, for example, tended to want to withhold rights from the veterans (e.g., suffrage) and generally held scorn for them because they were presumed poor and held responsible for their own poverty (Wilkinson, 2013a). The veterans also engaged in behaviors that were viewed as inappropriate, such as drinking alcohol, gambling, and solicitation of prostitutes.[4] Wilkinson highlights, for example,

> As long as the behavior of the men conformed to social expectations for quaint elderly soldiers, the local newspapers published proud accounts about the aging heroes. . . . However as the Pacific Branch population continued to grow, the behavior of certain of its residents repeatedly clashed with the increasingly temperance-minded bent of Santa Monica's elites.
>
> (2013a, p. 199)

While local residents worked to curtail veterans' behaviors through political actions like outlawing saloons in the city limits and gerrymandering school voting districts, the Home itself appears to have used various forms of nature, including the animal menagerie, as one tactic among many to civilize the veterans to the middle landscape of Santa Monica. As I demonstrate further, the Home's bird collection was more amenable to this task than the Home's monkey collection, as the monkeys, especially one named Joe, seemed to have resisted captivity. In so doing, Joe acted in ways counter to the Home's project of civilizing the veterans to the local community.

Making way for the "right" kinds of nature

In late nineteenth-century southern California, many existing spaces previously not subjected to development were cleared to make way for urbanization and supplanted with specific forms of human-constructed nature, including the Pacific Branch.[5] Indeed, an 1888 *Santa Monica Outlook* article describing changes to the city of Santa Monica notes,

> The sheep and plover have had to go. The former no longer mingle their melodious voices with the roaring surf, and where we once so successfully hunted the latter, are situated the most beautiful houses. Elegant mansions,

surrounded by handsome grounds, are seen where the squirrel and ground owl used to have every thing [sic] their own way.

(Unknown Author, 1888)

An undated photograph of the Home taken by Putnam and Valentine, who were photographers active in Los Angeles from 1898–1912, shows a neatly landscaped and lush green space (Putnam & Valentine, n.d.), reminiscent of the constructed public parks that began to dominate the American landscape during the latter part of the nineteenth century, thanks in part to the park planner Frederick Law Olmstead (Hanson, 2002) (see Figure 5.1). In his historical account of the development of the city of Santa Monica, Ingersoll similarly notes,

The buildings [of the Home] are all surrounded by carefully kept grounds, which are adorned with trees and flowers. This is one of the most beautifully arranged and kept parks in the country . . . making it a perennial garden of beauty.

(1908, p. 340)

Later in the text, he even refers to the Home as a "great attraction" because of its grounds (1908, p. 341), which resonates with narratives at the time describing public parks and zoos as attractions (Hanson, 2002). According to Wilkinson (2013a), surrounding community members actually used the Home as a rural park where they held meetings, picnics, and other events, though there is no evidence demonstrating they visited the animals at the Home's zoo, bird collection, or farm.

While not all public parks became associated with zoos, Hanson notes that the combination of lush green space with captive animals provided a more authentic experience observing nature that aligned with middle-class viewers' perspectives and also "further[ed] goals of educating working class and immigrant visitors to middle-class standards of behavior" (2002, p. 29). Although soldiers residing at the Pacific Branch hearkened from varying class backgrounds, because of their "immigrant" (i.e., they were not Santa Monica natives and many were born outside of the U.S.), institutionalized, invalid, and veteran status, they were actually viewed by the local community as poor by their own fault and morally inferior as a result of that poverty and their inappropriate behaviors (Wilkinson, 2013b). Taken together, this evidence suggests that one tactic the Home employed to socialize the soldiers to middle-class life in Santa Monica was through establishing a park-like atmosphere within the physical boundaries of the Home that included captive wildlife. In this way, specifically bounded forms of nature (e.g., socially constructed landscapes and particular kinds of captive animals) (Emel, Wilbert, & Wolch, 2002) became associated with good moral character.

Animals at the Home

Primary historical evidence documenting the presence of animals at the Home comes mainly from local newspaper articles, transaction log books, and a small number of other materials (e.g., Ingersoll, 1908; Soldiers' Home Booklet, 1900), which is unsurprising given that locating animals in the archive is fraught for a variety of reasons (Fudge, 2000, 2014; Kean, 2012; Rutherford, Thorpe, & Sandberg, 2016).[6] Due to the contested nature of constructing, creating, and telling animal histories, in the following analysis I also briefly turn to ethological studies in an attempt to "read" and "interpret" the archive not strictly from the perspective of the human writer but from that of the animal. Although the Home was certainly not a full-fledged zoo, a hodgepodge of animals populated what was referred to in the local papers as the Home "zoo" or "menagerie," including birds, monkeys, coyotes, and "dogs, rabbits, and other animals" (Unknown Author A, 1896, n.p.).[7] According to rather detailed logs of purchased items, the Home appears to have initially invested federal monies in caring for these animals without the expectation of return revenue, as opposed to the animals at the Home's farm (e.g., pigs, sheep, cows, and chickens). Evidence suggests, for example, that the farm initially turned a profit (Unknown Author A, 1901; Ingersoll, 1908), but by the fall of 1901, the "splendid herd of milch cows, together with some draught horses, was disposed of on Friday on competitive bids" (Unknown Author C, 1901, p. A15), and by 1904 the Home's sheep flock was also approved for sale given "the product not paying the cost of maintaining" (Board of Managers, 1904). In mid-1908, "because of the costliness due to combining cholera culture with swine raising" (Unknown Author, 1908, p. 118), the Home apparently ceased its hog production activities.

Initially, the Home appears to have invested federal dollars in maintaining the zoo. For example, the *Minutes of the Council of Administration Pertaining to Post Funds for the NVDHS* (1901–1913) at Sawtelle reveals multiple transactions to purchase monkey cages, birds, wire mesh, and canvas hoods for bird aviaries, repairs to cages, and consulting services related to caring for and classifying the birds: "Capt. S. J. Reber proposed the expenditure of twenty five (25) dollars to pay an expert to examine the plans of the proposed new aviary and advise as to this classification of the birds" (December 9, 1901) and "Capt. S. J. Reber proposed the expenditure of twelve hundred and fifty (1250) dollars for the construction of a new aviary and monkey cages, with room for the exhibit of curios" (December 31, 1901). Several entries specifically named particular bird species, one an eagle and another a macaw, both of which apparently died and were stuffed and mounted in 1903: "The Treasurer presented a statement of the proposed expenditure of forty ($40) dollars for wire netting for the dividing of the Eagle's cage into smaller compartments" (May 23, 1903) and "Major H.E. Hasse, Surgeon, moved the expenditure of $7.50 for the mounting of a macaw parrot" (November 6, 1905).

However, by 1902, the *Los Angeles Times* reported that the Home's zoo collection was to be sold, despite a 1901 article indicating that the construction of a new zoo and aviary was pending and would provide a permanent home for the newly acquired eagle and resident condor (Unknown Author, 1901a, p. 11):

The miscellaneous collection of animals constituting the Home zoo is doomed. Assistant Inspector-General Smith [sic] on his late official visit to this branch condemned the entire outfit, and as soon as purchasers can be found it will be finally disposed of. There are several very cute monkeys in the lot, chief of which is "Josephine," who has been the veterans' pet for eleven years.

(Unknown Author B, 1902, p. 9)

While it is not entirely clear why Smith condemned the zoo, it is possible that some of the animals residing there came to be too difficult and/or expensive to care for and did not serve the purposes the zoo was intended to serve (i.e, socializing the veterans to the middle landscape of the Santa Monica community in much the same way contemporary public parks and zoos in cities across America sought to socialize their urban residents). By the time Smith ordered the zoo's removal, for example, there had already been issues with at least one of the monkeys, despite the description of Josephine in the above passage as "cute." In 1896 the collection's largest monkey, Joe, was described as "hard to manage" because he began to show "an increasing disposition toward ugliness," and when enraged would "throw stones and other things at the members outside the wire netting," finally biting one of his keepers (Unknown Author B, 1896, p. 31). By the following year, Joe met a fate similar to that many of the Home's farm animals would face, as he was traded for a monkey that was expected to "furnish amusement for the members . . . without displaying the vicious instincts as Joe has done" (Unknown Author, 1896, p. 29). While it is difficult to "read" animal agency into any of these archival materials (Buller, 2014; Kean, 2012; Lorimer & Srinivasan, 2013; Rutherford, Thorpe, & Sandberg, 2016), given the animals' obvious captivity, disposability, and lack of control over their lives, Joe seemed to exercise the only form of agency he could by biting and throwing stones. Ethological studies, for example, demonstrate that complex social primates like monkeys would not find the Home zoo's meager accommodations (monkey cages) fulfilling, and in return would resist the doldrums of captive life (e.g., Hosey, 2005). Although it is reported that the veterans were amused by Joe's antics, including the throwing of stones at the wire mesh – "[he] has caused not a few of the veterans to forget their pains while yielding to the inclination to hearty laughter" – the monkey's unruly behavior did not conform to the Home's nor the surrounding community's expectations for "quaint elderly soldiers" (Wilkinson, 2013a, p. 199). That is, Joe did not emulate the kinds of model behavior expected of the veterans.

Birds and "fairy lakes"

Although Smith ordered the dispensation of the Home's zoo in 1902, multiple lines of evidence suggest that the bird collection was supported by federal funds at least until 1913 (Board of Managers, 1910; Driggs, 2011) and that the birds remained an interest of Home residents at least until 1918 (Unknown Author, 1918). A 1902 *Los Angeles Times* article reported, for example, that a new aviary "70 feet front by 30 feet deep, and its extreme height – at the dome – 30 feet" was to be constructed to replace the "shabby old zoo" and that,

Inside will be trees and shrubs in abundance, miniature grass plats, cool, fairy lakes where the warblers may splash and bathe at pleasure, and nesting places secure from intrusion, and, altogether, it is intended to make the aviary a pleasant as well as ornamental addition to the home.

(Unknown Author A, 1902, p. A15)

While evidence cannot fully explain the retention of the birds, attitudes about the bird collection expressed in local newspapers suggest that it functioned, at least in part, as a civilizing tactic. Indeed, the birds, who comprised many different species (Unknown Author B, 1903), were thought to be residing in "pleasant captivity" as they made music for the vets (Unknown Author B, 1903, p. 6), despite the fact that many of them had been "brought from far lands and strange countries" (Unknown Author B, 1903, p. 6) and could no longer interact with many members of their own kind, making it doubtful that their experiences were entirely pleasant (e.g., Dickens, Earle, & Romero, 2009).[8] However, as Grier (2006) notes, whether the birds enjoyed captivity or not, they remained popular pets during the early twentieth century and served as models for middle-class family life.

Unlike the descriptions of Joe's unruly behavior, local newspapers touted the caged birds as providing a source of genteel entertainment for the men through their sometimes "strange" and other times "old, familiar" songs (Unknown Author B, 1903, p. 6). "The Birds at Sawtelle," for example, contrasts the songs of the birds, "And we'll wager this is sweeter music to the ears of thee old comrades" to "when the bullets were singing at Antietam and Gettysburg," intimating the veterans lost a sort of personal tranquility during their service that they could potentially regain through contact with the Home's birds The author goes on to laud the veterans for their noble work, "It was not from desire that they fought, but for the love of their country along, and to preserve that country's integrity and honor," and their ability to "find happiness in the gentlest of pleasures – the song of a bird" (Unknown Author B, 1903, p. 6), which resonates with Wilkinson's (2013b) assertion that as long as the veterans' behavior conformed to the idealized expectations of the community, they were praised as heroes. In this particular case, it was not simply their service that earned them accolades, but their willingness to find solace with the captive birds and "the thousands of birds out and among the trees and shrubbery, coming and going at will" (Unknown Author B, 1903, p. 6). Such praise, therefore, emerged not from an absolute appreciation for the battles they fought, but out of a desire to turn the unrefined and "staunch soldiers" (Unknown Author B, 1903, p. 6) into a proper kind of Santa Monica civilian – one who is an honorable and tranquil comrade. The middle landscape, then, was embedded within the confines of the Home as the captive birds, wild birds, and aviary space became a tactic to civilize the veterans to the Santa Monica community.

Conclusion

This chapter has demonstrated how the Pacific Branch embedded what Hanson (2002) describes as the middle landscape concept within the space of the Home. This was accomplished through the construction of a public park used by local community members and through the installation of a small zoo and aviary that veterans apparently cared for and accessed for entertainment. Although historical evidence documenting the use of animals at the Home is limited, the Home's captive wild animals (especially the bird collection) appear to have been leveraged, at least in part, to socialize the veterans to the middle-class ideals of surrounding Santa Monica – a community that welcomed the veterans' pensions and resources, but not their gambling, drinking, solicitation of prostitutes, and, perhaps most importantly, assumed poverty (Wilkinson, 2013b). Although modern zoos and parks were cropping up in cities all over North America during the Victorian era, the current analysis uniquely illuminates typical spatio-temporal human–nature relations promoted within urbanizing cities during that time period. Here, the idealized interactions of public parks and zoos actually became mapped onto the private sphere of the Home, which the federal government and its local representatives tightly regulated. Given that the Home operated as a federal asylum for disabled veterans and was envisioned as an attraction for the Santa Monica community – making it wholly unlike a traditional domestic home – it is perhaps unsurprising that public ideals would interpenetrate this space. The Pacific Branch case nevertheless illustrates the complex ways in which: (1) public–private boundaries are blurred, especially when macro (e.g., federal) and micro (e.g., local) scales of government become involved in traditionally private affairs; and (2) human–nature relations are controlled and idealized in order to effectuate particular socioecological goals (i.e., using animals and/or other forms of nature as a tactic to civilize particular groups of people, even within the space of the home). While scholars have written quite extensively about the historical functions of zoos, aquariums, and wild animal parks (e.g., Anderson, 1995; Braverman, 2013; Rutherford, 2011), including exploring the role of local and regional politics in institution-making (Lloro-Bidart, 2015, 2016), little has been written about the history of spaces like the Home, which embraced the middle landscape concept at the local scale and within the private level of the Home.

Acknowledgments

I would like to thank my friend, Melissa Potter, for lending her expertise to me while I was conceptualizing and doing the archival research for this project. I'd also like to thank my colleague and friend, Connie Russell, for her constant support and guidance.

Notes

1 In his book-length treatment of the NHDVS, Kelly (1997) maintains that residents of the NHDVS "were not considered social deviants," and that "There was no sense, then, that the bodies and minds of Home residents required disciplining or rehabilitation to yield to bourgeois social norms" (1997, p. 166). Yet Kelly (1997) does not at all discuss the Pacific Branch in his book, which is arguably a unique case. Wilkinson's (2013a, 2013b) extensive research of the Pacific Branch demonstrates, for example, that "The increasingly temperance-minded town [of Santa Monica] took measures to curb what it considered to be the immoral behaviors of all of its visitors, including those of the growing number of Pacific Branch residents" (2013a, p. 199).

2 As Hanson notes, zoos also have colonial and imperial roots and have depended on the domination and exploitation of both animals and humans (e.g., Anderson, 1995; Chrulew, 2011; Lukasik, 2013; Ritvo, 1987). In invoking Hanson's (2002) "middle landscape" here, I do not wish to downplay the significance of power relationships in these spaces, but rather find the concept useful to explain the case of the Pacific Branch Soldiers' Home. Because the animal collection housed there was not a formal zoo and appeared, especially in the case of the bird collection, to be kept in order to provide enjoyment and socialization for the veterans who had physical and sustained contact with the animals (versus fleetingly gazing at them in the case of a more formal zoo), the middle landscape concept provides the analytical tools to understand these relationships not as colonial or imperial per se, but rather as civilizing and uplifting.

3 As Wilkinson (2013a) explains, Robert S. Baker's wife, Arcadia Bandini de Baker, was also wealthy, and when her husband died in 1894, she maintained the partnership he had established with John P. Jones.

4 Logue similarly highlights that the homes generally adopted "policies for paupers" in order to account for what was perceived as the veterans' moral failings. In order to circumvent the provision of services to just anyone and to ensure order, the homes adopted a military approach to organization, making continual efforts to control residents' behaviors (1992, pp. 413–416).

5 It should be noted here that it is also well-established that Indigenous peoples were forcibly removed from these lands, along with plants and animals. See Ingersoll (1908) for a historical account of these changes written contemporary to the operation of the Home.

6 While space limitations prohibit a detailed discussion of the methods and methodologies employed in this project, I conducted extensive searches through historical newspapers – online archives of the *Los Angeles Times* (1923–present), the *Santa Monica Evening Outlook* (1875–1936), the *Los Angeles Herald* (1873–1921), and a scrapbook of newspaper clippings (1894–1898) housed at the National Archives in Riverside, California – as well as the Sawtelle Disabled Veterans Home, Los Angeles Case Files, 1888–1933 (National Archives in Riverside, CA) and the Minutes of the Council of Administration Pertaining to Post Funds for the NVDHS at Sawtelle (National Archives in Riverside, CA). In these searches, I used specific animals as key words (i.e., "bird," "monkey," "rabbit," "coyote," "squirrel") in order to capture a wide variety of animals. I also consulted the archivist at the Santa Monica Historical Museum, who coincidentally was preparing an exhibit on the Home. She provided me information from a small booklet documenting daily life at the home and referred me to a descendent of Arcadia Bandini de Baker, Robert S. Baker's spouse, who took over his business dealings with John P. Jones after Robert S. Baker's death in 1894. After searching through their family's personal archival materials, they contacted me and indicated they had no information relevant to the current project.

7 A 1947 article in the Walnut Grower's Association Trade Magazine, "Diamond Walnut News," asserts that veterans living at the Home had family members residing in Tennessee or Mississippi send them eastern fox squirrels to serve as pets – or possibly food, since squirrels have been a popular item in southern cuisine and markers of southern identities (Tippen, 2014). According to Becker and Kimball (1947), who worked for the Los Angeles County Agricultural Commissioner's office and Los Angeles County, respectively, the squirrels were kept in cages for a time, but were turned loose when a hospital administrator found out about them and decided it was a poor use of federal resources to house and feed the squirrels. Since this article was published over four decades after the eastern fox squirrels were reported to have arrived at the Home, and given that I was unable to locate any other primary historical evidence to document the story, I do not include the eastern fox squirrels in the current analysis. In addition to the materials described previously, I also consulted with the Los Angeles County Agricultural Commissioner's office. While this search turned up information about the eastern fox squirrels contemporary to the "Diamond Walnut News" article, there was no

evidence supporting the connection between the Home and the eastern fox squirrels. It is worth noting that DNA haplotype analysis recently revealed that some eastern fox squirrel populations in California originated in the Mississippi Valley (Claytor, Muchilinski, & Torres, 2015). Furthermore, a 1905 *Santa Monica Outlook* article written by an unknown author lamented the presence of the "red squirrel" (a common name for the eastern fox squirrel) in southern California, suggesting that eastern fox squirrels had arrived by that time (Unknown Author, 1905). Yet neither piece of evidence irrefutably supports the Becker and Kimball (1947) story, pointing to the difficulty of piecing together historical animal geographies. See Lloro-Bidart (2017) for more regarding this case.

8 Hanson (2002) notes that many animals died during their journeys to U.S. zoos and circuses.

References

Anderson, K. (1995). Culture and nature at the Adelaide Zoo: At the frontiers of 'human geography'. *Transactions of the Institute of British Geographers, 20*(3), 275–294.

Becker, E. M., & Kimball, E. H. (1947). Walnut growers turn squirrel catchers. *Diamond Walnut News, 29*(3), 4–6.

Board of Managers. (1904). *Report of the board of managers of the national home for disabled volunteer soldiers for the fiscal year ended in June 30, 1904*. Washington, DC: Government Printing Office.

Board of Managers. (1910). *Report of the board of managers of the national home for disabled volunteer soldiers for the fiscal year ended in June 30, 1910*. Washington, DC: Government Printing Office.

Braverman, I. (2013). *Zooland: The institution of captivity*. Stanford, CA: Stanford University Press.

Buller, H. (2014). Animal geographies II: Methods. *Progress in Human Geography, 39*(3), 374–384.

Chrulew, M. (2011). Managing love and death at the zoo: The biopolitics of endangered species preservation. *Australian Humanities Review, 50*, 137–157.

Claytor, S. C., Muchilinski, A. C., & Torres, E. (2015). Multiple introductions of the eastern fox squirrel (*Sciurus niger*) in California. *Mitochondrial DNA, 26*(4), 583–592.

Dickens, M. J., Earle, K. A., & Romero, M. (2009). Initial transference of wild birds to captivity alters stress physiology. *General and Comparative Endocrinology, 160*(1), 76–83.

Driggs, J. (2011). Federal homemaking. *Strawberry Gazette, 7*, 1, 3.

Emel, J., Wilbert, C., & Wolch, J. (2002). Animal geographies. *Society & Animals, 10*(4), 407–412.

Fudge, E. (2000). Introduction to special edition: Reading animals. *Worldviews, 4*, 101–113.

Fudge, E. (2008). *Pets: The art of living*. Stocksfield, UK: Acumen.

Fudge, E. (2014). What was it like to be a cow? History and animal studies. In L. Kalof (Ed.), *The Oxford handbook of animal studies* (E Book) (n.p.). Oxford: Oxford Handbooks Online. doi: 10.1093/oxfordhb/9780199927142.013.28.

Grier, K. (2006). *Pets in America*. Chapel Hill, NC: University of North Carolina Press.

Hanson, E. (2002). *Animal attractions*. Princeton, NJ: Princeton University Press.

Hosey, G. R. (2005). How does the zoo environment affect the behaviour of captive primates? *Applied Animal Behaviour Science, 90*, 107–129.

Hovorka, A. (2015). The Gender, Place, and Culture Jan Monk Distinguished Annual Lecture: Feminism and animals: Exploring interspecies relations through intersectionality, performativity, and standpoint. *Gender, Place, & Culture: A Journal of Feminist Geography, 22*(1), 1–19.

Hribal, J. (2007). Animals, agency, and class: Writing the history of animals from below. *Human Ecology Review, 14*(1), 101–112.

Ingersoll, L. A. (1908). *Ingersoll's century history Santa Monica Bay cities.* Los Angeles, CA: L. A. Ingersoll.

Kean, H. (2012). Challenges for historians writing animal-human history: What's really enough? *Anthrozoös, 25,* S57–S72.

Kelly, P. J. (1997). *Creating a national home: Building the veterans' welfare state 1860–1900.* Cambridge, MA: Harvard University Press.

Lloro-Bidart, T. (2015). Neoliberal and disciplinary environmentality and 'sustainable seafood' consumption: Storying environmentally responsible action. *Environmental Education Research.* Advance online publication. doi: 10.1080/13504622.2015.1105198.

Lloro-Bidart, T. (2016). A feminist posthumanist political ecology of education for theorizing human-animal relations/relationships. *Environmental Education Research, 23*(1), 111–130.

Lloro-Bidart, T. (2017). When 'Angelino' squirrels don't eat nuts: A feminist posthumanist politics of consumption across southern California. *Gender, Place, & Culture: A Journal of Feminist Geography, 24*(6), 753–773.

Logue, L. M. (1992). Union veterans and their government: The effects of public policies on private lives. *Journal of Interdisciplinary History, 22*(3), 411–434.

Lorimer, J., & Srinivasan, K. (2013). Animal geographies. In N. C. Johnson, R. H. Schein, & J. Winders (Eds.), *The Wiley-Blackwell companion to cultural geography* (pp. 332–342). New York: John Wiley & Sons.

Lukasik, J. (2013). Embracing my escape from the zoo: A (sometimes) true account of my curricular inquiry. *Green Theory & Praxis Journal, 7*(1), 3–16.

Minutes of the Council of Administration Pertaining to Post Funds (1901–1913). Box 11, Bound Volume. National Home for Disabled Volunteer Soldiers. Pacific Branch (Sawtelle, California). National Archives at Riverside. Perris, CA. 7 September 2016.

Putnam, J. R., & Valentine, R. S. (Photographers). (No Date). Soldiers' Home, Views of Los Angeles, California. California Historical Society, CHS2013.1297.

Rader, K. A., & Cain, V.E.M. (2014). *Life on display: Revolutionizing U.S. museums in the twentieth century.* Chicago, IL: University of Chicago Press.

Ritvo, H. (1987). *The animal estate: The English and other creatures in the Victorian Age.* Cambridge, MA: Harvard University Press.

Rutherford, S. (2011). *Governing the wild: Ecotours of power.* Minneapolis, MN: University of Minnesota Press.

Rutherford, S., Thorpe, J., & Sandberg, L. A. (2016). Introduction: Methodological challenges. In J. Thorpe, S. Rutherford, & L. A. Sandberg (Eds.), *Methodological challenges in nature-culture and environmental history research* (pp. 26–36). New York: Routledge.

Sawtelle Disabled Veterans Home, Los Angeles Case Files (1888–1933). Boxes 1–8. National Home for Disabled Volunteer Soldiers. Pacific Branch (Sawtelle, California). National Archives at Riverside. Perris, CA. 7 September 2016.

Soldiers' Home Booklet (1900). Santa Monica History Museum Collection, 2016.24. Santa Monica History Museum, Santa Monica, CA.

Tippen, C. H. (2014). Squirrel, if you're so inclined. *Food, Culture, & Society, 17*(4), 555–570.

Unknown Author. (1888, January 18). A school boy takes a look ahead at Santa Monica. *Santa Monica Outlook*. Retrieved from http://digital.smpl.org/cdm/compoundobject/collection/outlook/id/37755/rec/2.

Unknown Author. (1891, October 16). Shot Abe: An editor kills the pet of the soldiers' home. *Los Angeles Herald*. Retrieved from https://cdnc.ucr.edu/cgi-bin/cdnc?a=d&d=LAH18911016.2.12&srpos=1&e=-----en-20-LAH-1-txt-txIN-An+Editor+Kills+the+Pet+of+the+Soldiers%e2%80%99+------1.

Unknown Author A. (1896, July 12). Soldiers' home: At the menagerie. *Los Angeles Times*. Retrieved from ProQuest Historical Newspapers: Los Angeles Times.

Unknown Author B. (1896, December 6). Soldiers' home: An ill-tempered ape assails and wounds a veteran. *Los Angeles Times*. Retrieved from ProQuest Historical Newspapers: Los Angeles Times.

Unknown Author A. (1901, April 14). Soldiers' home: Second in farm products. *Los Angeles Times*. Retrieved from ProQuest Historical Newspapers: Los Angeles Times.

Unknown Author B. (1901, August 11). Soldiers' home: Presented with an eagle. *Los Angeles Times*. Retrieved from ProQuest Historical Newspapers: Los Angeles Times.

Unknown Author C. (1901, October 27). Soldiers' home: Gen Brown inspects soldiers' home. *Los Angeles Times*. Retrieved from ProQuest Historical Newspapers: Los Angeles Times.

Unknown Author A. (1902, February 23). Soldiers' home: Birds about to sing. *Los Angeles Times*. Retrieved from ProQuest Historical Newspapers: Los Angeles Times.

Unknown Author B. (1902, July 20). Soldiers' home: Zoo must go. *Los Angeles Times*. Retrieved from ProQuest Historical Newspapers: Los Angeles Times.

Unknown Author A. (1903, November 15). Soldiers' home: Birds entertain veterans. *Los Angeles Times*. Retrieved from ProQuest Historical Newspapers: Los Angeles Times.

Unknown Author B. (1903, November 16). The birds at Sawtelle. *Los Angeles Times*. Retrieved from ProQuest Historical Newspapers: Los Angeles Times.

Unknown Author. (1905, May 23). Squirrel is a criminal: Little rodent declared to be a thief and a murderer by nature student. *Santa Monica Outlook*. Retrieved from http://digital.smpl.org/cdm/compoundobject/collection/outlook/id/28486/rec/139.

Unknown Author. (1908, June 5). Hear no more 'Oof, Oof, Oof': Banished are swine from the soldiers' home. *Los Angeles Times*. Retrieved from ProQuest Historical Newspapers: Los Angeles Times.

Unknown Author. (1911, September 10). The city of Sawtelle and the soldiers' home. *Los Angeles Times*. Retrieved from ProQuest Historical Newspapers: Los Angeles Times.

Unknown Author. (1918, December 22). Look out there, Teddy!: Humming bird [sic] eats from medicine dropper. *Los Angeles Times*. Retrieved from ProQuest Historical Newspapers: Los Angeles Times.

van de Hoek, R. J. (2005). National Soldiers' Home in West Los Angeles, California: Vernal Pools and Walnut Woodlands from 1890 to 2005. Retrieved from www.naturespeace.org/national.soldiers.home.htm.

Wilkinson, C. L. (2013a). The soldier's city: Sawtelle, California, 1897–1922. *Southern California Quarterly, 95*(2), 188–226.

Wilkinson, C. L. (2013b). *The veteran's in our midst: Disabled union veterans in west Los Angeles (1888–1914)*. (Unpublished master's thesis). California State University, Northridge, Northridge, CA.

Woods, M. (2000). Fantastic Mr. Fox? Representing animals in the hunting debate. In C. Philo & C. Wilbert (Eds.), *Animal spaces, beastly places: New geographies of human-animal relations* (pp. 182–202). London and New York: Routledge.

4 Shaking the ground

Histories of earthworms from Darwin to niche construction

Camilla Royle

Worms have played a more important part in the history of the world than most persons would at first suppose.

– Charles Darwin (1881)

The Darwin Centre in London's Natural History Museum contains 22 million animal specimens preserved in spirit (ethanol) and therefore known as the spirit collection.[1] The specimens available for public viewing – including bottled snakes, fish, octopuses, birds, rodents, chimps and even an eight-meter-long squid – represent just a small portion of this immense collection. Members of the public are able to get a close look, taking in the macabre spectacle of seeing various dead animals kept suspended and motionless in an odd assortment of sealed glass jars (Figure 4.1). Some of the creatures were collected, preserved and labeled by Charles Darwin himself. These include pickled fish brought back from his voyage on the *Beagle*, a five-year-long, round-the-world trip that cemented the naturalist's reputation and planted the seeds of his most significant ideas on evolution by natural selection.

The jars in the spirit collection are arranged according to a taxonomic system established by Carl Linnaeus in the 18th century. Indeed, the specimens in the public parts of the museum, as well as behind the scenes, are similarly ordered. Exhibits about mammals are in one part of the museum and those about insects are in another. The animals are therefore seen as distinct from their own habitats and also from other organisms with which they would interact in the wild.

The ways in which animals are arranged in museum exhibits can influence the way we think about them (Kalof, Zammit-Lucia and Kelly, 2011). By being invited to view dead animal specimens, we are encouraged to see animals as objects for our observation, which can, in turn, produce what Kay Anderson identifies as human exceptionalism – "the positing of a separation between active human subjects and passive non-human objects" (Anderson, 2014, p. 5). Furthermore, one aspect of this exceptionalism is that the practice of displaying animals as museum exhibits lends itself to an understanding of animal life as ahistorical. Human society might have changed since 19th-century explorers traveled the world searching for specimens, but, we are led to believe, the animals remain the same. This

Figure 4.1 Reptiles and fish in ethanol on public display in London's Natural History
      Museum

Source: taken by the author

ahistorical conception of animals is not only blind to the ways in which animals
have changed over time, but it would also seem to foreclose any discussion of how
humans and animals have influenced each other throughout history. As we will see,
in the case of Darwin's specimens, this is ironic indeed. It was Darwin who gave
us the idea that lifeforms have a history.

    In fairness, museums are often aware of the importance of how they represent
animals. Although they need to preserve animal specimens, the London Natural
History Museum also has a mandate to facilitate scientific research. The museum
often hosts visiting scientists who use parts of the collection in their work, and they
(more reluctantly) loan their specimens out to other labs. This tension between the
museum's desires to represent the animals of the past and to foster an understanding of living, breathing animal life was highlighted by the ongoing redevelopment
of the museum's central Hintze Hall. The project will involve removing Dippy, the
famous cast of a *Diplodocus* skeleton, and replacing it with a real blue whale

skeleton suspended from the ceiling ("a perfect symbol of . . . hope"), along with other new exhibits that tell "the dramatic story of evolution, diversity in the world today and our urgent role in the planet's future" (Natural History Museum, 2016).

Nevertheless, exhibits like the spirit collection demonstrate the resilience of the notion of humans as observers of passive nonhuman natures and how this is maintained and expressed in cultural practices. If humans are seen as dynamic while other animals are static, this risks enforcing a form of human exceptionalism whereby the agency of other animals is not fully accounted for. The capacity to shape history, it would seem, is reserved for humans alone. But is there an alternative way of doing animal geography that conceives of animals as historical agents in their relations with humans?

Animal geography and the placing of animals

In the last two decades, human geography has undergone something of a revival in its engagement with animals. These new animal geographies have been particularly important in challenging existing notions of subjectivity and asking what the way we relate to animals tells us about ourselves (Wolch and Emel, 1998; Wolch, 2002). These geographies have also done much to reinvigorate discussions of the constitutive role nonhuman animals have played in human societies (Philo and Wilbert, 2000). Rather than concentrating solely on the ways in which animals are represented by humans, therefore giving the impression "that animals are merely passive surfaces on to which human groups inscribe imaginings and orderings of all kinds," animal geographers have asked "how animals themselves may figure in these practices," making animal "agency" one of their key concerns (Philo and Wilbert, 2000, p. 5 and pp. 14–23). Animals, they say, can refuse to fit the binaries that humans have imposed on them. These accounts of the things that animals can do have gone alongside a more general turn toward seeing nonhuman natures as "chaotic," "eventful," "vital," "lively," and "feral" that is gaining influence within geography (Braun, 2009). Bees and humans can participate alongside each other (Bingham, 2006). And charismatic elephants captivate us with their appearance and behavior, triggering an emotional response and influencing our conservation priorities (Lorimer, 2007).

A key and related contribution of animal geography has been its refusal to treat the nonhuman nature as a background against which human history plays out, "a mute and stable background to the real business of politics," as Steve Hinchliffe puts it (2008). Such thinking is a challenge to notions that "nature" is a blank canvas prior to its being worked on by humans. Similarly, Donna Haraway is highly critical of the "productivist" logic that treats nature as a raw material that just exists before it is appropriated (Haraway, 1989, p. 13).

Animal geographers have also discussed the relationship between animals and spatiality. For Philo and Wilbert (2000), animals can be "placed" both in terms of being assigned to a place in a classification system, such as the Linnaean one mentioned previously, and by being more literally associated with particular physical places. These two forms of emplacement, conceptual and material, are closely linked. Scientific practices reinforce the idea that there is a "proper" place for animals. This process of placing animals is closely related to the ways in which nonhuman animals come to be differentiated from humans, with animals seen as belonging in different places and spaces (i.e., the outdoors, the wilderness) to those occupied by humans, although these boundaries are constantly contested (Philo and Wilbert, 2000, pp. 6–14).

If scientific norms such as taxonomy have served to establish a proper place for animals, then the presentation of animal specimens in a museum can similarly be thought of as a form of cultural practice. Rather than being simply expressions of ideas distinct from material reality, these practices of representing animals both emerge from and reinforce certain ways of treating nonhumans. Furthermore, as animals are often thought of as part of "nature," ways of seeing animals are closely related to understandings of the nonhuman world more broadly. Indeed, research on specific animals has been key to adding texture to environmental geography, opening the black box represented by the term "nature" (Wolch and Emel, 1998). So ideas about animals have real consequences for how humans relate to the rest of nature. Ideas matter.[2]

In the light of this, Anderson (2014) concludes her own discussion of human exceptionalism by discussing the stakes involved for environmental politics. She argues that assumptions of a passive, objectified nature are associated with the view that nature is malleable. If it is seen as a backdrop against which human activity takes place, it can become all too easy to assume that nonhuman nature can be manipulated in line with human desires. In what is now commonly referred to as the Anthropocene, a proposed geological epoch defined by the extent of human influence on Earth system processes, we are told that humans now have an unprecedented capacity to direct what happens to the rest of the biosphere. For example, Erle Ellis asserts that humans have become a great causal agent such that: "In moving toward a better Anthropocene, the environment will be what we make it" (quoted in Collard, Dempsey and Sundberg, 2015, p. 324).[3] Anderson points out the paradox here: renewed concerns over environmental threats in the 21st century have brought home how dependent human survival is on the rest of nature. However, "this threat also appears to be prompting a renewed commitment to the idea that humans possess a unique capacity to control our environment" (2014, p. 13).

So animal geography, as developed in the last two decades, has addressed, among other things, issues of animal subjectivity, their constitutive role in human societies and the role of scientific and other practices in representing animals, and the political implications of such practices. Therefore, this new animal geography provides a basis for an exploration of animals as historical agents and a discussion of the political consequences of not treating them as such.

However, as the editors of this volume remind us, it is easily forgotten that animals and other nonhumans change over time. Geography's accounts of animals have often focused on what animals do in the here and now. For example, Jamie Lorimer's work on nonhuman charisma (Lorimer, 2007, 2015) considers the fleeting effects of individual animals on the humans they come into contact with.

Relatively few have undertaken studies of animals in the past or addressed the extent to which nonhumans might be historicized.

This chapter explores the resources from biology that might enable a more historical understanding of animals (and other nonhuman species). It discusses Darwin's own work, particularly his writings on the "small agencies" of earthworms, before turning to more recent discussions around niche construction theory, which aims to extend the classical Darwinian approach by integrating an understanding of the way organisms change their environments through time to evolutionary theory. As we shall see, biology has its own history of engaging with the activities of animals, one where how to conceive of the historical activities of animals has become a source of heated debate both in the past and today.

Darwin on the action of worms

No discussion of animals and history would be complete if it did not mention Charles Darwin, who gave modern biology the idea that species undergo historical change. Before Darwin's theory of evolution by natural selection, life had been assumed to be typified by invariance and stability. According to biologist Ernst Mayr, Darwin therefore "introduced historicity into science" (Mayr, 2000).

Darwin was writing at a time when the foundations of biology were being shaken. In the 18th century, human exceptionalism had taken the form of Cartesian dualism, the idea, associated with René Descartes, that humans – uniquely among animals – possess a soul and/or the capacity for rational action. However, by the beginning of the 19th century, "the idea of an immaterial soul or mind was actively rejected" (Anderson, 2014, p. 6). This concept had given way to a more materialist approach rooted in the scientific practices of the day. This is not to say that human exceptionalism was dispensed with; rather, it persisted in a different form, with human distinctiveness thought of as having emerged from the physical differences between humans and other animals (Anderson, 2014, pp. 9–13). Therefore, Anderson argues, those who attack "Cartesian" dualism ought to note that they are dealing with a doctrine that has itself gone through historical changes. Furthermore, as Juanita Sundberg (2014) reminds us, the modern separation of nature and culture often referred to as "Cartesian," far from being a universal foundation of thought, was specific to Western societies.

Darwin was part of the attack on notions of a vital force that Anderson identifies. He repeatedly turned to the question of whether matter is animated by a force endowed on it from without or whether it contains an inherent vitality. As a young man observing pollen grains bursting under a microscope, he noted, "the matter inside seemed to have a self-activating power" (Desmond and Moore, 1991, p. 82). Indeed, this question enthralled the 19th-century world. Many of Darwin's contemporaries maintained that matter was essentially inert, with the vital force in living things given by God. However, some radicals, including Darwin's mentor Robert Grant, sought to explain liveliness in matter according to naturalistic principles without the need for celestial intervention.[4] This materialist and historical approach would be evident in Darwin's writings for the rest of his career.

Toward the end of his life, Darwin published an investigation into the "small agencies" of earthworms: *The Formation of Vegetable Mould through the Action of Worms with Observations on their Habits* (Darwin, 1881, p. 2). The book was a bestseller, with sales at the time actually rivaling those of his better known *On the Origin of Species* (Feller et al., 2003).

Devoted almost entirely to earthworms, the book includes the results of experiments on live worms that Darwin performed in his home in glass jars, observations of worms in the garden and notes on their historical activities. Darwin was interested in whether worms display evidence of "intelligence." He described how the animals responded to vibrations and appeared to feel "aroused," "surprised," or "distressed" by the presence of bright light (1881, p. 21). The possibility of worm intelligence was a key reason for the book's surprising popularity with the Victorian public. Worms had been considered ugly and useless, especially given their limited sensory abilities. The benefits of earthworms to gardeners were not well known even into the 20th century (Feller et al., 2003, p. 40).

It should be cautioned that Darwin hastily attributed intelligence to worms and that his writings would today be seen as needlessly anthropomorphic (Feller et al., 2003, p. 43). However, it is interesting that, given the modern association of "work" with human enterprise, Darwin refers repeatedly to the "work" carried out by the worms. For example, they are described as performing the work of drawing leaves into their burrows during the night (1881, p. 61). Darwin even explicitly compares the worms' actions to those of humans. He describes how, when a worm pulls a leaf into its burrow, it acts "in nearly the same manner as would a man" in solving this problem (1881, p. 312). Strikingly, in his concluding remarks he states that the earth has been "ploughed" by worms since long before humans invented the plough (p. 313). This suggests a perspective quite different to what Haraway criticizes as productivist logic.

Darwin described how worms tunnel into the soil and produce casts on the surface by ingesting a mixture of soil and vegetable matter and excreting it. For him it was fascinating to think that "the vegetable mould [topsoil] over the whole country has passed many times through, and will again pass many times through, the intestinal canals of worms" (1881, p. 4). Stones left on the surface of soil would, over decades, sink into the ground due to the actions of worms digging, bringing fresh soil to the surface in the form of excreted casts and creating tunnels underneath objects. For Darwin the whole of the topsoil is in "constant, though slow movement" (1881, p. 305). He tested this by setting out a "worm stone": a flat, round stone on the top of the soil with a hollow center containing an instrument for measuring how far the stone had sunk into the ground. A reconstruction of the experiment can be seen on the grounds of the family home at Down House in Kent, so the continued digging activities of the descendants of Darwin's worms can still be observed.

Darwin says that archaeologists ought to be thankful to worms, because they have preserved Roman ruins by covering them with earth. However, their activities are also destructive. Their steady burrowing has affected the Neolithic earthworks at Stonehenge, causing some of the outer stones to collapse and become partially buried (1881, pp. 154–158). The lives of worms are entwined with those of humans in other ways. As Darwin convincingly demonstrated, worms play a constitutive role in constructing human societies. They produce "wonderful" transformations in farmers' fields, turning rocky ground into fertile farmland.

In *Vibrant Matter*, Jane Bennett (2010) offers a re-reading of Darwin's book, addressing this role for worms in human societies as well as his comments on worm intelligence. Although Bennett says that worms are like us, she also departs slightly from Darwin's anthropocentrism in that human activity is not seen here as a yardstick against which other forms of agency are measured. For Bennett, agency is not a possession of humans but is distributed across a network. She notes how the way in which worms allow plants (including crops) to grow makes human action reliant on them. They should therefore be seen as members of a public in an interpretation of the political that is expanded beyond humans (pp. 103–104).

Darwin's views on worms suggest that, far from being hidden from contemporary observers, the activities of animals in the past are very much still observable today due to the effect creatures like worms have on the landscape. As evolutionary biologist Stephen Jay Gould (2007) points out, a book of observations of worms in the garden might be seen as a departure from work that had addressed grand questions – the origin of species, the evolution of humans – and a turn toward the mundane. However, Gould says that Darwin was actually pointing out something key to his whole way of thinking. Worms take small actions, an individual may not change very much in its lifetime, but when the actions of many worms are aggregated together the effect they can have is immense. Small, quantitative changes could trigger qualitative shifts, as evidenced by the collapsing stones at Stonehenge. In the same way, in evolutionary theory species might only change very slightly with each generation, but over millennia evolution can have a spectacular transformative effect. Failure to appreciate this had led to misunderstandings and even hostility toward evolution in Darwin's own time: "an . . . inability to sum up the effects of a continually recurrent cause, which has often retarded the progress of science" (Darwin, 1881, p. 6).

More recent investigations have described earthworms and other soil-dwelling invertebrates as ecosystem engineers – species that modify the resources available to other organisms by changing the physical, chemical or biological environment. As Lavelle et al. (2006) confirm, earthworms break up soil particles into smaller pieces, making the soil more amenable for crop plants to establish roots. They "dramatically change the structure and chemistry of the soils in which they live" by mixing organic material with the soil, burrowing and casting (Odling-Smee, Laland and Feldman, 2003, p. 11). Their physical effects can also be seen at much larger scales, for example, when the actions of worms on sloping environments erodes the soil and leads to soil creep (Lavelle et al., 2006, p. 8). However, Feller et al. (2003) suggest that Darwin's work on worms has been overlooked as agricultural scientists have become preoccupied with chemical inputs rather than soil composition or soil-dwelling organisms (2003, pp. 29 and 44). For these authors there is likely to be a revival of awareness and interest in the book in line with an increased interest in organic farming. However, caution is required where animals are viewed as providing services to humans. Where the input of nonhumans into agricultural processes has been recognized, various governments have been quick to try to incorporate their activity into a capitalist framework in the form of "natural capital" (Monbiot, 2014).

Recently, earthworms and their agency have again become a topic of debate among biologists. The theory of niche construction is proposed as an extension of Darwin's insights so that it accounts for the way in which organisms in the past, such as the worms, have created the environments that their descendants live in.

Niche construction: do organisms make their environments?

In October 2014, the scientific journal *Nature* ran a debate between two groups of biologists on the subject of whether evolutionary biology needs a rethink (Laland et al., 2014). Some, including Kevin Laland at the University of St Andrews, argued in favor of a shift in the central tenets of evolutionary theory. He says that theorists should incorporate an understanding of processes of niche construction, the ways in which living things construct their environment, and acknowledge the role this process plays in evolution. According to theories of niche construction, living things alter their own surroundings (and those of other organisms around them) and also expose their own offspring to these altered environments, which are sometimes referred to as their ecological inheritance. In this way they can be said to drive the evolutionary process by influencing the selection pressures on their own offspring. Therefore, niche construction represents a second route through which organisms pass on information to their descendants, by environmental modification as well as in the form of genes, which are passed down more directly to their offspring through reproduction. Niche construction, for its advocates, is as general a process within evolution as natural selection – and as significant. This second route "has been almost completely ignored in evolutionary biology" (Brandon and Antonovics, 1996, p. 176) in favor of a standard evolutionary theory which tends to play down the active role of organisms, treating them as passively responding to the pressures imposed by the environment (Odling-Smee, Laland and Feldman, 2003; Laland and Sterelny, 2006; Royle, 2017). Laland and his colleagues have sometimes used the term "agency," claiming on their website that organisms "exhibit active agency," as well as being "co-directors of their own evolution" and "driv[ing] environmental change" (see http://lalandlab.st-andrews.ac.uk/niche).

Laland's interlocutors, who say that all is well with evolutionary theory, agree that biology should deal with organism–environment relationships. However, they claim biology already does this. Reminding readers of Darwin's writings on earthworms, they point out that existing evolutionary theory already describes how organisms change their immediate surroundings and say that biologists have been studying this "for well over a century" (Laland et al., 2014, p. 163). However, although niche construction theorists concur with the theory of evolution by natural selection, their approach offers a subtle but important rethink of how evolution has generally been understood.

In conventional evolutionary theory after Darwin, individuals within a population will vary in their characteristics. Those with the characteristics most suited to a particular niche will be slightly more likely to survive and reproduce than their peers, meaning that their characteristics will be passed on to future generations. The best-adapted individuals are "selected" by the environment. Darwin was familiar with how this process might work due to his conversations with animal breeders, who literally select the most desirable birds for their own purposes and breed from them (see Desmond and Moore, 1991, pp. 425–430). Although natural selection is sometimes termed "survival of the fittest" (a term coined by Herbert Spencer), "fitness" in biology relates to how suited an organism is to its environment rather than being an inherent property (Mayr, 2000).[5]

Darwinian natural selection can therefore be seen as assuming that organisms solve a problem posed by the circumstances in which they live by evolving a response (Brandon and Antonovics, 1996). Or, in other words, they must fit into a pre-established template represented by the niche (Laland, Odling-Smee and Feldman, 2004). But organisms also modify their niche in multiple ways, both by choosing to live in a particular environment or by actively changing their environment (e.g., by burrowing, building nests or webs, or altering the chemical composition of soil). According to niche construction theorists, these are not incidental to evolution but play an important role.

The theory therefore treats organism and environment as "coevolving" or mutually influencing each other (Odling-Smee, Laland and Feldman, 2003, p. 2). Strikingly, some biologists have referred to the way organisms modify environments as "repeatable," "directional" and "systematic." Just as the persistent tunneling of the earthworms changes the landscape over time in a particular direction, the actions of organisms might also direct evolution (Laland et al., 2014, p. 162). This is not to say that nonhumans consciously act in pursuit of a particular goal; rather, it implies that they display goal-directed behavior to the extent that they have evolved to act in a particular way. A beaver building a dam under the influence of its genes might be said to act in a purposeful way even if it doesn't stop to think about why it needs a dam. However, the notion of purpose in evolutionary biology has been particularly controversial, likely to "make steam rise from an evolutionary biologist's ears," as *New Scientist* magazine put it (Holmes, 2013). Richard Dawkins, a popular science writer and niche construction sceptic, declined the magazine's request for comment and has, in the past, described the reasoning behind niche construction as "pernicious" (see Laland, Odling-Smee and Feldman, 2004). However, supporters of the theory contend that some phenomena in biology cannot be explained without it. As we shall see, earthworm physiology is one example of this.

The actions of worms have not only undermined buildings but have also played a role in earthworm evolution. Worms have modified the soil around them such that present-day individuals are exposed to a niche constructed by their ancestors over many generations. (Laland, Odling-Smee and Feldman, 2004). Earthworms evolved from aquatic animals. Tunneling and dragging of leaf litter below the surface has modified the soil to the extent that worms can regulate the water and salt balance in their bodies without needing to evolve internal kidneys like other terrestrial animals. They have therefore retained the nephridia or kidney-like structures of their freshwater relatives. In an aquatic environment, the nephridia serve to remove excess water from the animal's body; in worms the same organs extract water from the soil (Odling-Smee, Laland and Feldman, 2003). In short, worms have not so much adapted to their environment as refused to adapt, instead changing the environment to suit their physiology. Importantly, niche construction theorists argue that worm evolution cannot be adequately explained under the assumption that they simply adapt to a pre-existing niche. They need to make substantial changes to their environment in order to survive (Odling-Smee, Laland and Feldman, 2003, pp. 374–376). As Richard Lewontin has put it: "organisms do not adapt to their environments; they construct them out of the bits and pieces of the external world" (Lewontin, 1983, p. 280).

The origins of niche construction theory lie in a tradition of dialectical thinking in biology which is most closely associated with Richard Levins and Richard Lewontin and their 1985 book *The Dialectical Biologist*. Levins and Lewontin put forward an approach to biology influenced by Marxist dialectics, stating that organisms should be thought of as subjects of evolutionary processes as well as objects on which evolution acts (Levins and Lewontin, 1985, pp. 87–89; see also Lewontin, 1983). It is argued that the theory still demonstrates a dialectical sensibility. Perhaps its associations with these unconventional thinkers is one reason why niche construction theory has been controversial. Laland et al. have also suggested that biologists have felt the need to play down their differences, emphasizing natural selection at the expense of other processes that might be at work in evolution, in order to present a "united front" against anti-evolutionists (Laland et al., 2014).

It would be possible to consider the organism–environment relationship as one where two separate entities relate to each other in a simple back-and-forth relationship. Sometimes the metaphor of billiard balls colliding is invoked to describe such a relationship where one entity affects another from the outside without itself changing (Rose, 1997, p. 210). However, the more radical interpretation of niche construction challenges precisely the notion that there is such a prior separation between organism and environment (Royle, 2017; Vandermeer, 2008). Rather, by stressing the way organism and environment construct each other over time, the dialectical approach suggests that neither can exist independently of the other. So the theory presents a challenge to those approaches that begin with the notion that organism and environment are separate entities that then come to relate to each other. As Levins and Lewontin point out, organisms construct their environments at different spatial scales. For example, larger or more mobile organisms will often be able to construct environments at larger spatial scales than smaller or more sedentary ones can. Small organisms such as bacteria will be affected by processes such as Brownian motion that are irrelevant to larger ones (Levins and Lewontin, 1985, p. 104). These discussions imply a nonhuman production of scale that is sometimes sidelined in discussions of scale within human geography.

Furthermore, for Brandon and Antonovics (1996, p. 174), niche construction is also inherently about how organisms change over time and how this is understood in relation to the changes they make to their environments. They point out that from an atemporal point of view, aspects of the environment can be measured independently of the organism that inhabits it. For example, one might measure the temperature, pH or numbers of a microbe living in soil. However, from an evolutionary point of view the environment is both cause and effect of evolutionary processes so cannot be seen as simply independent. So the temporal dimension, as well as geographical space and scale, needs to be taken into account in order to understand how organism and environment mutually construct each other.

Conclusion

It is easy to forget that animals have a history. This chapter has argued that the tendency to treat humans as historical actors and nonhumans as ahistorical is reinforced by cultural practices such as the ways in which animals are presented in natural history museums. Neglecting to historicize animals is one aspect of a human exceptionalism that treats them as part of a passive "nature" that serves as a background to human activity. Biology, since Darwin established the theory of evolution by natural selection, has offered an alternative vision of living things, including animals as dynamic rather than static, a sensibility evident in his final book on earthworms. If more processes such as niche construction are incorporated into evolutionary theory, it may also provide a fuller account of the ways in which organisms and environments do not just evolve but coevolve. While accounts of animals from animal geography have sometimes tended toward considering individual animals and the ways in which humans relate to them, when niche construction biologists consider the actions of earthworms, they are talking about a species in general and change occurring over long timescales. Worms modify the soil over decades; the evolution of their physiology has taken place over millions of years. Their activity may be slow, but over time it can have dramatic effects, including contributing to soils that human society relies on for agriculture. For niche construction biologists, animals do not merely have histories; they also play a role in shaping history.

Some attempts to account for the role of worms have involved trying to assign monetary values to their activities. However, if there are problematic ways of dealing with nonhuman agency, there are also dangers in seeing humans as the sole possessors of agency over the rest of the biosphere. Taking an historical approach to animal geography reminds us that humans do not manipulate a passive and malleable earth; we act on an earth that is already being continually churned by worms.

Acknowledgments

I would like to thank Alex Loftus and my fellow Geography students at King's College for their advice in writing this chapter.

Notes

1 I would like to thank the biologists who agreed to be interviewed for this research and the staff at the Natural History Museum, London and Down House, Kent.
2 See, for example, Jason Moore's (2015, chapter 8) discussion of the ways in which practices of measuring and categorizing nonhuman natures cannot be seen as distinct from the commodification and appropriation of those natures.
3 Although this is only one interpretation of the meaning of the Anthropocene – see Royle (2016) for my own views.
4 As Darwin's biographers point out, Grant's view fitted well with the political views of the liberal intellectual circles with which Darwin associated. It suggested that people were capable of self-improvement, providing an analogy that suited the emerging industrial class when old institutions such as the church and aristocracy were losing some of their authority (Desmond and Moore, 1991, p. 223).
5 Fitness is defined as how successful an organism is at reproducing.

References

Anderson, K. (2014). Mind over matter? On decentring the human in human geography. *Cultural Geographies*, 21(1), 3–18.

Bennett, J. (2010). *Vibrant Matter: A Political Ecology of Things*. Durham and London: Duke University Press.

Bingham, N. (2006). Bees, butterflies, and bacteria: Biotechnology and the politics of nonhuman friendship. *Environment and Planning A*, 38, 483–498.

Brandon, R. N. & Antonovics, J. (1996). The coevolution of organism and environment. In R.N Brandon (Ed.), *Concepts and Methods in Evolutionary Biology* (pp. 161–178). Cambridge: Cambridge University Press.

Braun, B. (2009). Nature. In N. Castree, D. Demeritt, D. Liverman & B. Rhoads (Eds.), *A Companion to Environmental Geography* (pp. 19–36). Oxford: Wiley-Blackwell.

Collard, R., Dempsey, J. & Sundberg, J. (2015). A manifesto for abundant futures. *Annals of the Association of American Geographers*, 105(2), 322–330.

Darwin, C. (1881). *The Formation of Vegetable Mould through the Action of Worms, with Observations on Their Habits*. London: John Murray.

Desmond, A. & Moore, J. (1991). *Darwin*. London: Penguin.

Feller, C., Brown, G. G., Blanchart, E., Deleporte, P. & Chernyanskii, S. S. (2003). Charles Darwin, earthworms and the natural sciences: Various lessons from past to future. *Agriculture, Ecosystems and Environment*, 99, 29–49.

Gould, S. J. (2007 [1982]). A worm for a century and all seasons. In P. McGarr & S. Rose (Eds.), *The Richness of Life: The Essential Stephen Jay Gould* (pp. 155–165). London: Vintage.

Haraway, D. (1989). *Primate Visions: Gender, Race and Nature in the World of Modern Science*. New York and London: Routledge.

Hinchliffe, S. (2008). Reconstituting nature conservation: Towards a careful political ecology. *Geoforum*, 39(1), 88–97.

Holmes, B. (2013, 12 October). Life's purpose. *New Scientist*, 33–35.

Kalof, L., Zammit-Lucia, J. & Kelly, J. R. (2011). The meaning of animal portraiture in a museum setting: Implications for conservation. *Organization and Environment*, 24(2), 150–174.

Laland, K., Odling-Smee, J. & Feldman, M. (2004). Causing a commotion: Niche construction: Do the changes that organisms make to their habitats transform evolution and influence natural selection? *Nature*, 429, 609.

Laland, K. & Sterelny, K. (2006). Perspective: Seven reasons (not) to neglect niche construction. *Evolution*, 60(9), 1751–1762.

Laland, K. et al. (2014). Does evolutionary theory need a rethink? *Nature*, 514, 161–164.

Lavelle, P., Decaëns, T., Aubert, M., Barot, S., Blouin, M., Bureau, F., Margerie, P., Mora, P. & Rossi, J.-P. (2006). Soil invertebrates and ecosystem services. *European Journal of Soil Biology*, 42, 3–15.

Levins, R. & Lewontin, R. C. (1985). *The Dialectical Biologist*. Cambridge, MA: Harvard University Press.

Lewontin, R. C. (1983). Gene, organism and environment. In D. S. Bendall (Ed.), *Evolution from Molecules to Men* (pp. 273–285). Cambridge: Cambridge University Press.

Lorimer, J. (2007). Nonhuman charisma. *Environment and Planning D: Society and Space*, 25, 911–932.

Lorimer, J. (2015). *Wildlife in the Anthropocene*. Minneapolis, MN: University of Minnesota Press.

Mayr, E. (2000, 1 July). Darwin's influence on modern thought. *Scientific American*, www.scientificamerican.com/article/darwins-influence-on-modern-thought/.

Monbiot, G. (2014, 24 July). Put a price on nature? We must stop this neoliberal road to ruin. *Guardian*, www.theguardian.com/environment/georgemonbiot/2014/jul/24/price-nature-neoliberal-capital-road-ruin.

Moore, J. W. (2015). *Capitalism in the Web of Life*. London and New York: Verso.

Natural History Museum. (2016, 19 December). Blue whale: Summer 2016, www.nhm.ac.uk/press-office/press-releases/blue-whale-summer-2017.html.

Odling-Smee, F. J., Laland, K. & Feldman, M. W. (2003). *Niche Construction: A Neglected Process in Evolution*. Princeton, NJ: Princeton University Press.

Philo, C. & Wilbert, C. (Eds.). (2000). *Animal Spaces, Beastly Places: New Geographies of Human-Animal Relations*. London and New York: Routledge.

Rose, S. (1997). *Lifelines: Life beyond the Gene*. London: Vintage.

Royle, C. (2016). Marxism and the Anthropocene. *International Socialism*, 151, 63–84.

Royle, C. (2017). Complexity, dynamism and agency: How can dialectical biology inform geography? *Antipode*, 49(5), 1427–1445.

Sundberg, J. (2014). Decolonizing posthumanist geographies. *Cultural Geographies*, 21(1), 33–47.

Vandermeer, J. (2008). The niche construction paradigm in ecological time. *Ecological Modelling*, 214, 385–390.

Wolch, J. R. (2002). Anima urbis. *Progress in Human Geography*, 26(6), 721–742.

Wolch, J. R. & Emel, J. (Eds.). (1998). *Animal Geographies: Place, Politics, and Identity in the Nature-Culture Borderlands*. London and New York: Verso.

Part II

The city – historical animals in and out of sight

5   Zoöpolis

Jennifer Wolch

[W]ithout the recognition that the city is of and within the environment, the wilderness of the wolf and the moose, the nature that most of us think of as natural cannot survive, and our own survival on the planet will come into question.[1]

Introduction

Urbanization in the West was based historically on a notion of progress rooted in the conquest and exploitation of nature by culture. The moral compass of city builders pointed toward the virtues of reason, progress, and profit, leaving wild lands and wild things – as well as people deemed to be wild or "savage" – beyond the scope of their reckoning. Today, the logic of capitalist urbanization still proceeds without regard to nonhuman animal life, except as cash-on-the-hoof headed for slaughter on the "disassembly" line or commodities used to further the cycle of accumulation.[2] Development may be slowed by laws protecting endangered species, but you will rarely see the bulldozers stopping to gently place rabbits or reptiles out of harm's way.

Paralleling this disregard for nonhuman life, you will find no mention of animals in contemporary urban theory, whose lexicon reveals a deep-seated anthropocentrism. In mainstream theory, urbanization transforms "empty" land through a process called "development" to produce "improved land," whose developers are exhorted (at least in neoclassical theory) to dedicate it to the "highest and best use." Such language is perverse: wildlands are not "empty" but teeming with nonhuman life; "development" involves a thorough denaturalization of the environment; "improved land" is invariably impoverished in terms of soil quality, drainage, and vegetation; and judgements of "highest and best use" reflect profit-centered values and the interests of humans alone, ignoring not only wild or feral animals but captives such as pets, lab animals, and livestock, who live and die in urban space shared with people. Marxian and feminist varieties of urban theory are equally anthropocentric.[3]

Our theories and practices of urbanization have contributed to disastrous ecological effects. Wildlife habitat is being destroyed at record rates as the urban front advances worldwide, driven in the First World by suburbanization and edge-city development, and in the Second and Third Worlds by pursuit of a "catching-up" development model that produces vast rural to urban migration flows and sprawling squatter landscapes.[4] Entire ecosystems and species are threatened, while individual animals in search of food and/or water must risk entry into urban areas, where they encounter people, vehicles, and other dangers. The explosion of urban pet populations has not only polluted urban waterways but led to mass killings of dogs and cats. Isolation of urban people from the domestic animals they eat has distanced them from the horrors and ecological harms of factory farming, and the escalating destruction of rangelands and forests driven by the market's efforts to create/satisfy a lust for meat. For most free creatures, as well as staggering numbers of captives such as pets and livestock, cities imply suffering, death, or extinction.

The aim of this paper is to foreground an urban theory that takes nonhumans seriously. Such a theory needs to address questions about (1) how urbanization of the natural environment impacts animals, and what global, national, and locality-specific political-economic and cultural forces drive modes of urbanization that are most threatening to animals; (2) how and why city residents react to the presence of animals in their midst, why attitudes may shift with new forms of urbanization, and what this means for animals; (3) how both city-building practices and human attitudes and behaviors together define the capacity of urban ecologies to support nonhuman life; and (4) how the planning/policy-making activities of the state, environmental design practices, and political struggles have emerged to slow the rate of violence toward animals witnessed under contemporary capitalist urbanization. In the first part, I clarify what I mean by "humans" and "animals," and provide a series of arguments suggesting that a trans-species urban theory is necessary to the development of an eco-socialist, feminist, anti-racist urban praxis. Then, in the second part, I argue that current considerations of animals and people in the capitalist city (based on US experience) are strictly limited, and suggest that a trans-species urban theory must be grounded in contemporary theoretical debates regarding urbanization, nature and culture, ecology, and urban environmental action.

Why animals matter (even in cities)

The rationale for considering animals in the context of urban environmentalism is not transparent. Urban environmental issues traditionally center around the pollution of the city conceived as human habitat, not animal habitat. Thus the various wings of the urban progressive environmental movement have avoided thinking about nonhumans and have left the ethical as well as pragmatic ecological, political, and economic questions regarding animals to be dealt with by those involved in the defense of endangered species or animal welfare. Such a division of labor privileges the rare and the tame, and ignores the lives and living spaces of the large number and variety of animals who dwell in cities. In this section, I argue that even common, everyday animals should matter.

The human–animal divide: a definition

At the outset, it is imperative to clarify what we mean when we talk about "animals" or "nonhumans" on the one hand, and "people" or "humans" on the other.

Where does one draw the line between the two, and upon what criteria? In many parts of the world beliefs in transmogrification or transmigration of souls provide a basis for beliefs in human–animal continuity (or even coincidence). But in the Western world animals have for many centuries been defined as fundamentally different and ontologically separate from humans, and although explicit criteria for establishing human–animal difference have changed over time, all such criteria routinely use humans as the standard for judgement. The concern is, can animals do what humans do? rather than, can humans do what animals do? Thus judged, animals are inferior beings. The Darwinian revolution declared a fundamental continuity between the species, but standing below humans on the evolutionary scale, animals could still be readily separated from people, objectified and used instrumentally for food, clothes, transportation, company, or spare body parts.

Agreement about the human–animal divide has recently collapsed. Critiques of post-Enlightenment science,[5] greater understanding of animal thinking and capabilities, and studies of human biology and behavior emphasizing human–animal similarities have all rendered claims about human uniqueness deeply suspect. Debates about the human–animal divide have also raged as a result of sociobiological discourses about the biological bases for human social organization and behavior, and feminist and anti-racist arguments about the social bases for human differences claimed to be biological. Long-held beliefs in the human as social subject and the animal as biological object have thus been destabilized.

My position on the human–animal divide is that animals as well as people socially construct their worlds and influence each other's worlds. The resulting "animal constructs are likely to be markedly different from ours but may be no less real."[6] Animals have their own realities, their own worldviews; in short, they are *subjects*, not objects. This position is rarely reflected in eco-socialist, feminist and anti-racist practice, however. Developed in direct opposition to a capitalist system riddled by divisions of class, race/ethnicity, and gender, and deeply destructive of nature, such practice ignores some sorts of animals altogether (for example, pets, livestock) or has embedded animals within holistic and/or anthropocentric conceptions of the environment and therefore avoided the question of animal subjectivity.[7] Thus, in most forms of progressive environmentalism, animals have been objectified and/or backgrounded.

Thinking like a bat: the question of animal standpoints

The recovery of animal subjectivity implies an ethical and political obligation to redefine the urban problematic and to consider strategies for urban praxis from the standpoints of animals. Granting animals subjectivity at a theoretical, conceptual level is a first step. Even this first step is apt to be hotly contested by human social groups who have been marginalized and devalued by claims that they are "closer to animals" and hence less intelligent, worthy, or evolved than Anglo-European white males. It may also run counter to those who interpret the granting of subjectivity as synonymous with a granting of rights and object either to rights-type arguments in general or to animal rights specifically.[8] But a far more difficult step must be taken if the revalorization of animal subjectivity is to be meaningful in terms of day-to-day practice. We not only have to "think like a mountain" but also to "think like a bat," somehow overcoming Nagel's classic objection that because bat sonar is not similar to any human sense, it is humanly impossible to answer a question such as "what is it like to be a bat?" or, more generally, "what is it like to be an animal?"[9]

But is it impossible to think like a bat? There is a parallel here with the problems raised by standpoint (or multipositionality) theories. Standpoint theories assert that a variety of individual human differences (such as race, class, or gender) so strongly shape experience and thus interpretations of the world that a single position essentializes and silences difference, and fails to challenge power relations. In the extreme, such polyvocality leads to a nihilistic relativism and a paralysis of political action. But the response cannot be to return to practices of radical exclusion and denial of difference. Instead, we must recognize that individual humans are embedded in social relations and networks with people similar or different upon whom their welfare depends.[10] This realization allows for a recognition of kinship but also of difference, since identities are defined through seeing that we are similar to, and different from, related others. And through everyday interaction and concerted practice, and using what Haraway terms a "cyborg vision" that allows "partial, locatable, critical knowledge sustaining the possibility of webs of connection called solidarity,"[11] we can embrace kinship as well as difference and encourage the emergence of an ethic of respect and mutuality, caring and friendship.[12]

The webs of kinships and difference that shape individual identity involve both humans and animals. This is reasonably easy to accept in the abstract (that is, humans depend upon a rich ecology of animal organisms). But there is also a large volume of archeological, paleoanthropological, and psychological evidence suggesting that concrete interactions and interdependence with animal others are indispensable to the development of human cognition, identity, and consciousness, and to a maturity that accepts ambiguity, difference, and lack of control.[13] In short, animals are not only "good to think" (to borrow a phrase from Lévi-Strauss) but indispensable to learning how to think in the first place, and how to relate to other people.

Who are the relevant animal others? I argue that many sorts of animals matter, including domestic animals. Clearly, domestication has profoundly altered the intelligence, senses, and life ways of creatures such as dogs, cows, sheep, and horses so as to drastically diminish their otherness; so denaturalized, they have come to be seen as part of human culture. But wild animals have been appropriated and denaturalized by people too. This is evidenced by the myriad ways wildlife is commercialized (in both embodied and disembodied forms) and incorporated into material culture. And like domestic animals, wild animals can be profoundly impacted by human actions, often leading to significant behavioral adaptations. Ultimately, the division between wild and domestic must be seen as a permeable social construct; it may be better to conceive of a *matrix* of animals who vary with respect to the extent of physical or behavioral modification due to human intervention, and types of interaction with people.

Our ontological dependency on animals seems to have characterized us as a species since the Pleistocene. Human needs for dietary protein, desires for spiritual inspiration and companionship, and the ever-present possibility of ending up as somebody's dinner required thinking like an animal. This aspect of animal contribution to human development can be used as an (anthropomorphic) argument in defense of wildlife conservation or pet keeping. But my concern is how human dependency on animals was played out in terms of the patterns of human–animal interactions it precipitated. Specifically, did ontological dependency on animals create an interspecific ethic of caring and webs of friendship? Without resurrecting a 1990s version of the Noble Savage – an essentialized indigenous person living in spiritual and material harmony with nature – it is clear that for most of (pre) history, people ate wild animals, tamed them, and kept them captive, but also respected them as kin, friends, teachers, spirits, or gods. Their value lay both in their similarities with and differences from humans. Not coincidentally, most wild animal habitats were also sustained.

Re-enchanting the city: an agenda to bring the animals back in

How can animals play their integral role in human ontology today, thereby helping to foster ethical responses and political practices engendered by the recognition of human–animal kinship and difference? Most critically, how can such responses and practices possibly develop in places where everyday interaction with so many kinds of animals has been eliminated? Most people now live in such places, namely cities. Cities are perceived as so human-dominated that they become naturalized as just another part of the ecosystem, that is, the human habitat. In the West, many of us interact with or experience animals only by keeping captives of a restricted variety or eating "food" animals sliced into steak, chop, and roast. We get a sense of wild animals only by watching "Wild Kingdom" reruns or going to Sea World to see the latest in a long string of short-lived "Shamus."[14] In our apparent mastery of urban nature, we are seemingly protected from all nature's dangers but chance losing any sense of wonder and awe for the nonhuman world. The loss of both the humility and the dignity of risk results in a widespread belief in the banality of day-to-day survival. This belief is deeply damaging to class, gender, and North–South relations as well as to nature.[15]

To allow for the emergence of an ethic, practice, and politics of caring for animals and nature, we need to renaturalize cities and invite the animals back in, and in the process re-enchant the city.[16] I call this renaturalized, re-enchanted city *zoöpolis*. The reintegration of people with animals and nature in zoöpolis can provide urban dwellers with the local, situated, everyday knowledge of animal life required to grasp animal standpoints or ways of being in the world, to interact with them accordingly in particular contexts, and to motivate political action necessary to protect their autonomy as subjects and their life spaces. Such knowledge would stimulate a thorough rethinking of a wide range of urban daily life practices: not only animal regulation and control practices, but landscaping, development rates and design, roadway and transportation decisions, use of energy, industrial toxics, and bioengineering – in short, all practices that impact animals and nature in its diverse forms (climate, plant life, landforms, and so on). And, at the most personal level, we might rethink eating habits, since factory farms are so environmentally destructive *in situ*, and the Western meat habit radically increases the rate at which wild habitats are converted to agricultural land worldwide (to say nothing of how one feels about eating cows, pigs, chickens, or fishes once they are embraced as kin).

While based in everyday practice like the bioregional paradigm, the renaturalization or zoöpolis model differs in including animals and nature in the metropolis rather than relying on an anti-urban spatial fix like small-scale communalism. It also accepts the reality of global interdependence rather than opting for autarky. Moreover, unlike deep ecological visions epistemically tied to a psychologized individualism and lacking in political-economic critique, urban renaturalization is motivated not only by a conviction that animals are central to human ontology in ways that enable the development of webs of kinship and caring with animal subjects, but that our alienation from animals results from specific political-economic structures, social relations, and institutions operative at several spatial scales. Such structures, relations, and institutions will not magically change once individuals recognize animal subjectivity, but will only be altered through political engagement and struggle against oppression based on class, race, gender, and species.

Beyond the city, the zoöpolis model serves as a powerful curb on the contradictory and colonizing environmental politics of the West as practiced both in the West itself and as inflicted on other parts of the world. For example, wildlife reserves are vital to prevent species extinction. But because they are "out there," remote from urban life, reserves can do nothing to alter entrenched modes of economic organization and associated consumption practices that hinge on continual growth and make reserves necessary in the first place. The only modes of life that the reserves change are those of subsistence peoples, who suddenly find themselves alienated from their traditional economic base and further immiserated. But an interspecific ethic of caring replaces dominionism to create urban regions where animals are not incarcerated, killed, or sent off to live in wildlife prisons, but instead are valued neighbors and partners in survival. This ethic links urban residents with peoples elsewhere in the world who have evolved ways of both surviving and sustaining the forests, streams, and diversity of animal lives, and enjoins their participation in the struggle. The Western myth of a pristine Arcadian wilderness, imposed with imperial impunity on those places held hostage to the International Monetary Fund and the World Bank in league with powerful international environmental organizations, is trumped by a post-colonial politics and practice that begins at home with animals in the city.

Ways of thinking animals in the city

An agenda for renaturalizing the city and bringing animals back in should be developed with an awareness of the impacts of urbanization on animals in the capitalist city, how urban residents think about and behave toward animal life, the ecological adaptations made by animals to urban conditions, and current practices and politics arising around urban animals. Studies that address these topics are primarily grounded in empiricist social science and wildlife biology. The challenge of trans-species urban theory is to develop a framework informed by social theory. The goal is to understand capitalist urbanization in a globalizing economy and what it means for animal life; how and why patterns of human–animal interactions change over time and space; urban animal ecology as science, social discourse, and political economy; and trans-species urban practice shaped by managerial plans and grassroots activism. Figure 5.1 lays out a metatheoretical heuristic

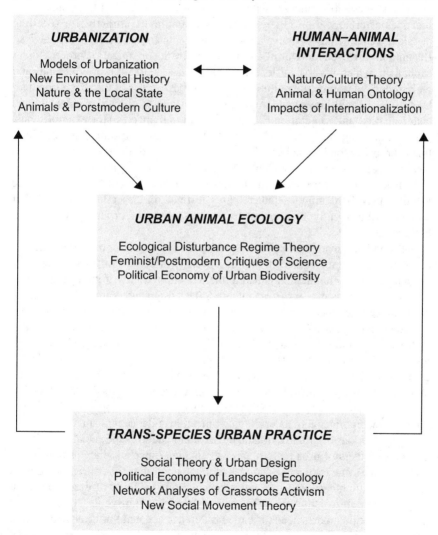

Figure 5.1 Conceptual framework for linking the disparate discourses of the trans-species urban problematic

device that links together the disparate discourses of the trans-species urban problematic. This device does not seek to privilege a particular theoretical perspective, but rather highlights multiple sources of inspiration that may be fruitful in theory development.

Animal town: urbanization, environmental change, and animal life chances

The city is built to accommodate humans and their pursuits, yet a subaltern "animal town" inevitably emerges with urban growth. This animal town shapes the practices of urbanization in key ways (for example, by attracting or repelling people/development in certain places, or influencing animal exclusion strategies). But animals are even more profoundly affected by the urbanization process under capitalism, which involves extensive denaturalization of rural or wild lands and widespread environmental pollution. The most basic types of urban environmental change are well-known and involve soils, hydrology, climate, ambient air and water quality, and vegetation.[17] Some wild animal species (for example, rats, pigeons, cockroaches) adapt to and/or thrive in cities. But others are unable to find appropriate food or shelter, adapt to urban climate, air quality, or hydrological changes, or tolerate contact with people. Captives, of course, are mostly restricted to homes, yards, or purpose-built quarters such as feed lots or labs, but even the health of pets, feral animals, and creatures destined for dissecting trays or dinner tables can be negatively affected by various forms of urban environmental pollution.

Metropolitan development also creates spatially extensive, patchy landscapes and extreme habitat fragmentation that especially affects wildlife. Some animals can adapt to such fragmentation and to the human proximity it implies, but more commonly animals die *in situ* or migrate to less fragmented areas. If movement corridors between habitat patches are cut off, species extinction can result as fragmentation intensifies, due to declining habitat patch size,[18] deleterious edge effects,[19] distance or isolation effects, and related shifts in community ecology.[20] Where fragmentation leads to the loss of large predators, remaining species may proliferate, degrade the environment, and threaten the viability of other forms of wildlife. Weedy, opportunistic, and/or exotic species may also invade, to similar effect.

Such accounts of urban environmental change and habitat fragmentation are not typically incorporated into theories of urbanization under capitalism. For example, most explanations of urbanization do not explicitly address the social or political-economic drivers of urban environmental change, especially habitat fragmentation.[21] By the same token, most studies of urban environments restrict themselves to the scientific measurement of environmental-quality shifts or describe habitat fragmentation in isolation from the social dynamics that drive it.[22] This suggests that urbanization models need to be reconsidered to account for the environmental as well as political-economic bases of urbanization, the range of institutional forces acting on the urban environment, and the cultural processes that background nature in the city.

Efforts to theoretically link urban and environmental change are at the heart of the new environmental history, which reorients ideas about urbanization by illustrating how environmental exploitation and disturbance underpin the history of cities, and how thinking about nature as an actor (rather than a passive object to be acted upon) can help us understand the course of urbanization. Contemporary urbanization, linked to global labor, capital, and commodity flows, is simultaneously rooted in exploitation of natural "resources" (including wildlife, domestic, and other sorts of animals) and actively transforms regional landscapes and the possibilities for animal life – although not always in the manner desired or expected, due to nature's agency. Revisiting neo-Marxian theories of the local state as well as neo-Weberian concepts of urban managerialism to analyze relations between nature and the local state could illuminate the structural and institutional contexts of, for example, habitat loss/degradation. One obvious starting place is growth machine theory, since it focuses on the influence of rentiers on the local state apparatus and local politics;[23] another is the critique of urban planning as part of the modernist project of control and domination of others (human as well as nonhuman) through rationalist city building and policing of urban interactions and human/animal proximities in the name of human health and welfare.[24] Finally, urban cultural studies may help us understand how the aesthetics of urban built environments deepen the distanciation between animals and people. For instance, Wilson demonstrates how urban simulacra such as zoos and wildlife parks have increasingly mediated human experience of animal life.[25] Real live animals can actually come to be seen as less than authentic since the terms of authenticity have been so thoroughly redefined. The distanciation of wild animals has simultaneously stimulated the elaboration of a romanticized wildness used as a means to peddle consumer goods, sell real estate, and sustain the capital accumulation process, reinforcing urban expansion and environmental degradation.[26]

Reckoning with the beast: human interactions with urban animals

The everyday behavior of urban residents also influences the possibilities for urban animal life. The question of human relations with animals in the city has been tackled by empirical researchers armed with behavioral models, who posit that, through their behavior, people make cities more or less attractive to animals (for example, human pest management and animal control practices, urban design, provision of food and water for feral animals and/or wildlife). These behaviors, in turn, rest on underlying values and attitudes toward animals. In such values-attitudes-behavior frameworks, resident responses are rooted in cultural beliefs about animals, but also in the behavior of animals themselves – their destructiveness, charisma and charm, and, less frequently, their ecological benefits.

Attitudes toward animals have been characterized on the basis of survey research and the development of attitudinal typologies.[27] Findings suggest that urbanization increases both distanciation from nature and concern for animal welfare. Kellert, for example, found that urban residents were less apt to hold utilitarian attitudes, were more likely to have moralistic and humanistic attitudes, suggesting that they were concerned for the ethical treatment of animals, and were focused on individual animals such as pets and popular wildlife species.[28] Urban residents of large cities were more supportive of protecting endangered species; less in favor of shooting or trapping predators to control damage to livestock; more apt to be opposed to hunting; and supportive of allocating additional public resources for programs to increase wildlife in cities. Domestic and attractive animals were most preferred, while animals known to cause human property damage or inflict injury were among the least preferred.

Conventional wisdom characterizes the responses of urban residents and institutions to local animals in two ways: (1) as "pests," who are implicitly granted agency in affecting the urban environment, given the social or economic costs they impose; or (2) as objectified "pets," who provide companionship, an aesthetic amenity to property owners, or recreational opportunities such as bird-watching and feeding wildlife.[29] Almost no systematic research, however, has been conducted on urban residents' behavior toward the wild or unfamiliar animals they encounter or how behavior is shaped by space or by class, patriarchy, or social constructions of race/ethnicity. Moreover, the behavior of urban institutions involved in urban wildlife management or animal regulation/control has yet to be explored.[30]

How can we gain a deeper understanding of human interactions with the city's animals? The insights from wider debates in nature/culture theory are most instructive and help put behavioral research in proper context.[31] Increasingly, nature/culture theorizing converges on the conviction that the Western nature/culture dualism, a variant of the more fundamental division between object and subject, is artificial and deeply destructive of Earth's diverse life-forms. It validates a theory and practice of human/nature relations that backgrounds human dependency on nature. Hyperseparating nature from culture encourages its colonization and domination. The nature/culture dualism also incorporates nature into culture, denying its subjectivity and giving it solely instrumental value. By homogenizing and disembodying nature, it becomes possible to ignore the consequences of human activity such as urbanization, industrial production, and agro-industrialization on specific creatures and their terrains. This helps trigger what O'Connor terms the "second contradiction of capitalism," that is, the destruction of the means of production via the process of capital accumulation itself.[32]

The place-specific version of the nature/culture dualism is the city/country divide; as that place historically emblematic of human culture, the city seeks to exclude all remnants of the country from its midst, especially wild animals. As we have already seen, the radical exclusion of most animals from everyday urban life may disrupt development of human consciousness and identity, and prevent the emergence of interspecific webs of friendship and concern. This argument filters through several variants of radical ecophilosophy. In some versions, the centrality of "wild" animals is emphasized, while the potential of tamer animals, more common in cities but often genetically colonized, commodified, and/or neotenized, is questioned. In other versions, the wild/tame distinction in fostering human–animal bonds is minimized, but the progressive loss of interspecific contact and thus understanding is mourned.[33] Corporeal identity may also become increasingly destabilized as understandings of human embodiment traditionally derived through direct experience of live animal bodies/subjects evaporates or is radically transformed. Thus what we now require are theoretical treatments explicating how the deeply ingrained dualism between city (culture) and country (nature), as it is played out ontologically, shapes human–animal interactions in the city.

The ahistorical and placeless values-attitudes-behavior models also miss the role of social and political-economic context on urban values and attitudes toward animals. Yet such values and attitudes are apt to evolve in response to place-specific situations and local contextual shifts resulting from nonlocal dynamics, for example, the rapid internationalization of urban economies. Deepening global competition threatens to stimulate a hardening of attitudes toward animal exploitation and habitat destruction in an international "race to the bottom" regarding environmental/animal protections. Moreover, globalization sharply reveals the fact that understandings of nature in the West are insufficient to grasp the range of relationships between people and animals in diverse global cities fed by international migrant flows from places where nature/culture relations are radically different. Variations on the theme of colonization are being played back onto the colonizers; in the context of internationalization, complex questions arise concerning how both colonially imposed, indigenous, and hybrid meanings and practices are being diffused back into the West. Also, given globalization-generated international migration flows to urban regions, we need to query the role of diverse cultural norms regarding animals in the racialization of immigrant groups and spread of nativism in the West. Urban practices that appear to be linked to immigrant racialization involve animal sacrifice (for example, Santeria) and eating animals traditionally considered in Western culture as household companions.

An urban bestiary: animal ecologies in the city

The recognition that many animals coexist with people in cities and the management implications of shared urban space have spurred the nascent field of urban animal ecology. Grounded in biological field studies and heavily management-oriented, studies of urban animal life focus on wildlife species; there are very few ecological studies of urban companion or feral animals.[34] Most studies tend to be highly species- and place-specific. Only a small number of urban species have been scrutinized, typically in response to human-perceived problems, risk of species endangerment, or their "charismatic" character.

Ecological theory has moved away from holism and equilibrium notions toward a recognition that processes of environmental disturbance, uncertainty, and risk cause ecosystems and populations to continually shift over certain ranges varying with site and scale.[35] This suggests the utility of reconceptualizing cities as ecological disturbance regimes rather than ecological sacrifice zones whose integrity has been irrevocably violated. In order to fully appreciate the permeability of the city/country divide, the heterogeneity and variable patchiness of urban habitats and the possibilities (rather than impossibilities) for urban animal life must be more fully incorporated into ecological analyses. This in turn could inform decisions concerning prospective land-use changes (such as suburban densification or down-zoning, landscaping schemes, transportation corridor design) and indicate how they might influence individual animals and faunal assemblages in terms of stress levels, morbidity and mortality, mobility and access to multiple sources of food and shelter, reproductive success, and exposure to predation.

Scientific urban animal ecology is grounded in instrumental rationality and oriented toward environmental control, perhaps more than other branches of ecology since it is largely applications driven. The effort by preeminent ecologist Michael Soulé to frame a response to the postmodern reinvention of nature, however, demonstrates the penetration into ecology of feminist and postmodern critiques of modernist science.[36] Hayles, for instance, argues that our understanding of nature is mediated by the embodied interactivity of observer and observed, and the positionality (gender, class, race, species) of the observer.[37] Animals, for example, construct different worlds through their embodied interactions with it (that is, how their sensory and intellectual capabilities result in their world-views). And although some models may be more or less adequate interpretations of nature, the question of how positionality determines the models proposed, tested, and interpreted must always remain open. At a minimum, such thinking calls for self-reflexivity in ecological research on urban animals and ecological tool-kits augmented by rich ethnographic accounts of animals, personal narratives of nonscientific observers, and folklore.

Finally, scientific urban animal ecology is not practiced in a vacuum. Rather, like any other scientific pursuit, it is strongly shaped by motives of research sponsors (especially the state), those who use research products (such as planners), and ideologies of researchers themselves. Building on the field of science studies, claims of scientific ecology must thus be interrogated to expose the political economy of urban animal ecology and biodiversity analysis. How are studies of urban animals framed, and from whose perspective? What motivates them in the first place – developer proposals, hunter lobbies, environmental/animal rights organizations? Sorting out such questions requires not only evaluation of the technical merits of urban wildlife studies, but also analysis of how they are framed by epistemological and discursive traditions in scientific ecology and embedded in larger social and political-economic contexts.

Redesigning nature's metropolis: from managerialism to grassroots action

A nascent trans-species urban practice, as yet poorly documented and under-theorized, has appeared in many US cities. This practice involves numerous actors, including a variety of federal, state, and local bureaucracies, planners, and managers, and urban grassroots animal/environmental activists. In varying measure, the goals of such practice include altering the nature of interactions between people and animals in the city, creating minimum-impact urban environmental designs, changing everyday practices of the local state (wildlife managers and urban planners), and more forcefully defending the interests of urban animal life.

Wildlife managers and pest-control firms increasingly face local demands for alternatives to extermination-oriented animal-control policies. In the wildlife area, approaches were initially driven by local protests against conventional practices such as culling; now managers are more apt to consider in advance resident reactions to management alternatives and to adopt participatory approaches to decision-making in order to avoid opposition campaigns. Typically, alternative management strategies require education of urban residents to increase knowledge and understanding of, and respect for, wild animal neighbors, and to underscore how domestic animals may harm or be harmed by wildlife. There are limits to educational approaches, however, stimulating some jurisdictions to enact regulatory controls on common residential architectures, building maintenance, garbage storage, fencing, landscaping, and companion-animal keeping that are detrimental to wildlife.

Wild animals were never a focus of urban and regional planning. Nor were other kinds of animals, despite the fact that a large proportion of homes in North America and Europe shelter domestic animals. This is not surprising given the historic location of planning within the development-driven local state apparatus. Since the passage of the US Endangered Species Act (ESA) in 1973, however, planners have been forced to grapple with the impact of human activities on threatened/ endangered species. To reduce the impact of urbanization on threatened/ endangered animals, planners have adopted such land-use tools as zoning (including urban limit lines and wildlife overlay zones), public/nonprofit land acquisition, transfer of development rights (TDR), environmental impact statements (EIS), and wildlife impact/habitat conservation linkage fees.[38] None of these tools is without severe and well-known technical, political, and economic problems, stimulating the development of approaches such as habitat conservation plans (HCPs) – regional landscape-scale planning efforts to avoid the fragmentation inherent in project-by-project planning and local zoning control.[39]

Despite the ESA, minimum-impact planning for urban wildlife has not been a priority for either architects or urban planners. Wildlife-oriented residential landscape architecture remains uncommon. Most examples are new developments (as opposed to retrofits), sited at the urban fringe, planned for low densities, and thus oriented for upper-income residents only. Many are merely ploys to enhance real-estate profits by providing home-buyers, steeped in an anti-urban ideology of suburban living emphasizing proximity to "the outdoors," with an extra "amenity" in the form of proximity to wild animals' bodies. Planning practice routinely defines other less attractive locations which host animals (dead or alive), such as slaughterhouses and factory farms, as "noxious" land uses and isolates them from urban residents to protect their sensibilities and the public health.

Wildlife considerations are also largely absent from the US progressive architecture/ planning agenda, as are concerns for captives such as pets or livestock. The 1980s "costs of sprawl" debate made no mention of wildlife habitat, and the adherents to the so-called new urbanism and sustainable cities movements of the 1990s rarely define sustainability in relation to animals. The new urbanism emphasizes sustainability through high density and mixed-use urban development, but remains strictly anthropocentric in perspective. Although more explicitly ecocentric, the sustainable cities movement aims to reduce human impacts on the natural environment through environmentally sound systems of solid-waste treatment, energy production, transportation, housing, and so on, and the development of urban agriculture capable of supporting local residents.[40] But while such approaches have long-term benefits for all living things, the sustainable cities literature pays little attention to questions of animals per se.[41]

Everyday practices of urban planners, landscape architects, and urban designers shape normative expectations and practical possibilities for human–animal interactions. But their practices do not reflect desires to enrich or facilitate interactions between people and animals through design, nor have they been assessed from this perspective. Even companion animals are ignored; despite the fact that there are more US households with companion animals than children, such animals remain invisible to architects and planners. What explains this anthropocentrism on the part of urban design and architectural professions? Social theories of urban design and professional practice could be used to better understand the anthropocentric production of urban space and place. Cuff, for example, explains the quotidian behavior of architects as part of a collective, interactive social process conditioned by institutional contexts including the local state and developer clients; not surprisingly, design outcomes reflect the growth orientation of contemporary urbanism.[42] More broadly, Evernden argues that planning and design professionals are constrained by the larger culture's insistence on rationality and order and the radical exclusion of animals from the city.[43] The look of the city as created by planners and architects, dominated by standardized design forms such as the suburban tract house surrounded by a manicured, fenced lawn, reflects the deep-seated need to protect the domain of human control by excluding weeds, dirt, and – by extension – nature itself.

Environmental designers drawing on conservation biology and landscape ecology have more actively engaged the question of how to design new metropolitan landscapes for animals and people than have planners or architects.[44] At the regional level, wildlife corridor plans or reserve networks are in vogue.[45] Wildlife networks and corridors are meant to link "mainland" habitats beyond the urban fringe, achieve overall landscape connectivity to protect gene pools, and provide habitat for animals with small home ranges.[46] Can corridors protect and reintegrate animals in the metropolis? Corridor planning is a recent development, and we need case-specific political-economic analyses of corridor plans to answer this question. Preliminary experience suggests that at best large-scale corridors can offer vital protection to gravely threatened keystone species and thus a variety of other animals, while small-scale corridors can be an excellent urban design strategy for allowing common small animals, insects, and birds to share urban living space with people. However, grand corridor proposals can degrade into an amenity for urban recreationists (since they often win taxpayers' support only if justified on recreational rather than habitat-conservation grounds). At worst, corridors may become a collaborationist strategy that merely smooths a pathway for urban real-estate development into wilderness areas.

A growing number of urban grassroots struggles revolves around the protection of specific wild animals or animal populations, and around the preservation of urban wetlands, forests, and other wildlife habitat due to their importance to wildlife. Also, growing awareness of companion-animal wants and desires has stimulated grassroots efforts to create specially designed spaces for pets in the city, such as dog parks.[47] But we have very little systematic information about what catalyzes such grassroots trans-species urban practices or about the connections between such struggles and other forms of local eco/animal activism. It is not clear if grassroots struggles around animals in the city are linked organizationally either to larger-scale environmental activism or green politics, or to traditional national animal welfare organizations, suggesting the need for mapping exercises and organizational network analyses. Ephemeral and limited case-study information suggests that political action around urban animals can expose deep divisions within environmentalism and the animal welfare establishment. These divisions mirror the broader political splits between mainstream environmentalism and the environmental justice movement, between animal rights organizations and environmentalists, and between groups with animal rights and groups with animal welfare orientations. For example, many mainstream groups only pay lip service (if that) to social justice issues, and so many activists of color continue to consider traditional environmental priorities such as wildlands and wildlife – especially in cities – as at best a frivolous obsession of affluent white suburban environmentalists, and at worst reflective of pervasive elitism and racism. Local struggles around wildlife issues can also expose the philosophical split between holistic environmental groups and individualist animal rights activists; for example, such conflicts often arise over proposals to kill feral animals in order to protect native species and ecosystem fragments. And reformist animal welfare organizations such as urban humane societies, concerned primarily with companion animals and often financially dependent on the local state, may be wary of siding with animal rights/liberation groups critical not only of state policies but also the standard practices of the humane societies themselves.[48]

The rise of organizations and informal groups acting to preserve animal habitat in the city, change management policies, and protect individual animals indicates a shift in everyday thinking about the positionality of animals. If such a shift is underway, why and why now? One possibility is that ecocentric environmental ethics and especially animal rights thinking, with its parallels between racism, sexism, and "speciesism," have permeated popular consciousness and stimulated new social movements around urban animals. Other avenues of explanation may open up by theorizing trans-species movements within the broader context of new social movement theory, which points to these movements' consumption-related focus; grassroots, localist, and anti-state nature; and linkages to the formation of new sociocultural identities necessitated by the postmodern condition and contemporary capitalism.[49] Viewed through the lens of new social movement theory, struggles to resist incursions of capital into urban wildlife habitat or defend the interests of animals in the city could be contextualized within larger social and political-economic dynamics as they alter forms of activism and change individual-level priorities for political action. Such an exercise might even reveal that new social movements around animals transcend both production and consumption-related concerns, reflecting instead a desire among some people to span the human–animal divide by extending networks of caring and friendship to nonhuman others.

Toward zoöpolis

Zoöpolis presents both challenges and opportunities for those committed to eco-socialist, feminist, and anti-racist urban futures. At one level, the challenge is to overcome deep divisions in theoretical thinking about nonhumans and their place in the human moral universe. Perhaps more crucial is the challenge of political practice, where purity of theory gives way to a more situated ethics, coalition building, and formation of strategic alliances. Can progressive urban environmentalism build a bridge to those people struggling around questions of urban animals, just as reds have reached out to greens, greens to feminists, feminists to those fighting racism? In time- and place-specific contexts where real linkages are forged, the range of potential alliances is apt to be great, extending from groups with substantial overlap with progressive environmental thinking to those whose communalities are more tenuous and whose focuses are more parochial. Making common cause on specific efforts to fight toxics, promote recycling, or shape air-quality management plans with grassroots groups whose raison d'être is urban wildlife, pets, or farm animal welfare may be difficult. The potential to expand and strengthen the movement is significant, however, and should not be overlooked.

The discourse of zoöpolis creates a space to initiate outreach, conversation, and collaboration in these borderlands of environmental action. Zoöpolis invites a critique of contemporary urbanization from the standpoints of animals but also from the perspective of people, who together with animals suffer from urban pollution and habitat degradation and who are denied the experience of animal kinship and otherness so vital to their well-being. Rejecting alienated theme-park models of human interaction with animals in the city, zoöpolis instead asks for a future in which animals and nature would no longer be incarcerated beyond the reach of our everyday lives, leaving us with only cartoons to heal the wounds of their absence. In a city re-enchanted by the animal kingdom, the once-solid Enchanted Kingdom might just melt into air.

Acknowledgments

A slightly longer version of this chapter appears in *Capitalism, Nature, Socialism* 7, 1996, pp. 2–48. I am grateful to Guilford Press for allowing it to be reproduced here.

Notes

1 Daniel B. Botkin, *Discordant Harmonies: A New Ecology for the Twenty-First Century*, New York: Oxford University Press, 1990, p. 167.

2 Such commodified animals include those providing city dwellers with opportunities for "nature consumption" and a vast array of captive and companion animals sold for profit.

3 For exceptions, see Ted Benton, *Natural Relations: Ecology, Animal Rights and Social Justice*, London: Verso Books, 1993; Barbara Noske, *Humans and Other Animals*, London: Unwin Hyman, 1989.

4 Maria Mies and Vandana Shiva, *Ecofeminism*, London: Zed Books, 1993.

5 For example, Lynda Birke and Ruth Hubbard, eds., *Reinventing Biology: Respect for Life and the Creation of Knowledge*, Bloomington and Indianapolis, IN: Indiana University Press, 1995.

6 Noske, *Humans and Other Animals*, p. 158; for similar perspectives, see also Donna Haraway, *Simians, Cyborgs, and Women: The Reinvention of Nature*, New York: Routledge, 1991; Val Plumwood, *Feminism and the Mastery of Nature*, London: Routledge, 1993, and, from the perspective of a biologist, Donald Griffin, *Animal Thinking*, Cambridge, MA: Harvard University Press, 1984.

7 Progressive environmental practice has conceptualized "the environment" as a scientifically defined system; as "natural resources" to be protected for human use; or as an active but unitary subject to be respected as an independent force with inherent value. The first two approaches are anthropocentric; the ecocentric third approach, common to several strands of green thought, is an improvement, but its ecological holism backgrounds interspecific difference among animals (human and nonhuman) as well as the difference between animate and inanimate nature.

8 A recovery of the animal subject does not imply that animals have rights, although the rights argument does hinge on the conviction that animals are subjects of a life; see Tom Regan, *The Case for Animal Rights*, Berkeley, CA: University of California Press, 1986.

9 Thomas Nagel, "What Is It Like to Be a Bat?," *The Philosophical Review* 83, 1974.

10 This argument follows those by Plumwood, *Feminism and the Mastery of Nature*. See also Jessica Benjamin, *The Bonds of Love: Psychoanalysis, Feminism and the Problem of Domination*, London: Virago, 1988; Jean Grimshaw, *Philosophy and Feminist Thinking*, Minneapolis, MN: University of Minnesota Press, 1986.

11 Donna Haraway, "Situated Knowledges: The Science Question in Feminism and the Privilege of Partial Perspective," in *Simians, Cyborgs, and Women: The Reinvention of Nature*, New York: Routledge, 1991, p. 191.

12 This in no way precludes self-defense against animals such as predators, parasites, or micro-organisms that threaten to harm people.

13 This evidence has perhaps most extensively been marshaled by Paul Shepard, *Thinking Animals: Animals and the Development of Human Intelligence*, New York: Viking Press, 1978; Paul Shepard, *Nature and Madness*, San Francisco, CA: Sierra Club Books, 1982; and, most recently, Paul Shepard, *The Others*, Washington, DC: Earth Island Press, 1996.

14 "Shamu" was the name used for a series of killer whales who performed in a major U.S. marine theme park.

15 Mies and Shiva, *Ecofeminism*.

16 As highlighted in the following section, there are many animals that do, in fact, inhabit urban areas. But most are uninvited, and many are actively expelled or exterminated. Moreover, animals have been largely excluded from our *understanding* of cities and urbanism.

17 Ann Whiston Sprin, *The Granite Garden: Urban Nature and Human Design*, New York: Basic Books, 1984; Michael Hough, *City Form and Natural Process*, New York: Routledge, 1995.

18 Otto H. Frankel and Michael E. Soulé, *Conservation and Evolution*, London: Cambridge University Press, 1981; Michael E. Gilpin and Ilkka Hanski, eds., *Metapopulation Dynamics: Empirical and Theoretical Investigations*, New York: Academic Press, 1991.

19 Michael E. Soulé, "Land Use Planning and Wildlife Maintenance: Guidelines for Conserving Wildlife in an Urban Landscape," *Journal of the American Planning Association* 57, 1991.

20 Mark L. Shaffer, "Minimum Population Sizes for Species Conservation," *BioScience* 31, 1981.

21 See, for example, Michael Dear and Allen J. Scott, *Urbanization and Urban Planning in Capitalist Society*, London: Methuen, 1981.

22 An example is Ian Laurie, ed., *Nature in Cities*, New York: Wiley, 1979.

23 John R. Logan and Harvey L. Molotch, *Urban Fortunes: The Political Economy of Place*, Berkeley, CA: University of California Press, 1987.

24 Elizabeth Wilson, *The Sphinx in the City: Urban Life, the Control of Disorder, and Women*, Berkeley: University of California Press, 1991; Christine M. Boyer, *Dreaming the Rational City: The Myth of American City Planning*, Cambridge, MA: MIT Press, 1983; Chris Philo, "Animals, Geography and the City: Notes on Inclusions and Exclusions," *Environment & Planning D: Society and Space* 13, 1995.

25 Alexander Wilson, *The Culture of Nature: North American Landscapes from Disneyland to the Exxon Valdez*, Cambridge, MA: Blackwell Books, 1992.

26 Gary Snyder, *The Practice of the Wild*, San Francisco, CA: North Point Press, 1990.

27 See the three-part study by Stephen R. Kellert, *Public Attitudes toward Critical Wildlife and Natural Habitat Issues, Phase I*, US Department of Interior, Fish and Wildlife Service, 1979; *Activities of the American Public Relating to Animals, Phase II*, US Department of Interior, Fish and Wildlife Service, 1980; and, co-authored with Joyce Berry, *Knowledge, Affection and Basic Attitudes toward Animals in American Society, Phase III*, US Department of Interior, Fish and Wildlife Service, 1980.

28 Stephen R. Kellert, "Urban Americans' Perceptions of Animals and the Natural Environment," *Urban Ecology* 8, 1984.

29 David A. King, Jody L. White, and William W. Shaw, "Influence of Urban Wildlife Habitats on the Value of Residential Properties," in L. W. Adams and D. L. Leedy, eds., *Wildlife Conservation in Metropolitan Environments*, National Institute for Urban Wildlife, 1991, pp. 165–169; William W. Shaw, J. Mangun, and R. Lyons, "Residential Enjoyment of Wildlife Resources by Americans," *Leisure Sciences* 7, 1985.

30 For an exception, see William W. Shaw and Vashti Supplee, "Wildlife Conservation in a Rapidly Expanding Metropolitan Area: Informational, Institutional and Economic Constraints and Solutions," in L. W. Adams and D. L. Leedy, eds., *Integrating Man and Nature in the Metropolitan Environment*, National Institute of Urban Wildlife, 1987, pp. 191–198.

31 Donna Haraway, *Primate Visions: Gender, Race, and Nature in the World of Modern Science*, New York: Routledge, 1989; Neil Evernden, *The Social Creation of Nature*, Baltimore, MD: Johns Hopkins University Press, 1992; Plumwood, *Feminism and the Mastery of Nature*.

32 James O'Connor, "Capitalism, Nature, Socialism: A Theoretical Introduction," *Capitalism, Nature, Socialism* 1, 1988.

33 Paul Shepard, "Our Animal Friends," in S. R. Kellert and E. O. Wilson, eds., *The Biophilia Hypothesis*, Washington, DC: Island Press, 1993, pp. 275–300, stresses the wild, while others are more inclusive, such as Noske, *Humans and Other Animals*; Karen Davis, "Thinking Like a Chicken: Farm Animals and the Feminine Connection," in Carol J. Adams and Josephine Donovan, eds., *Animals and Women: Feminist Theoretical Explorations*, Durham, NC and London: Duke University Press, 1995, pp. 192–212.

34 For exceptions, see Alan M. Beck, *The Ecology of Stray Dogs: A Study of Free-Ranging Urban Animals*, Baltimore, MD: York Press, 1974; C. Haspel and R. E. Calhoun, "Activity Patterns of Free-Ranging Cats in Brooklyn, New York," *Journal of Mammalogy* 74, 1993.

35 S.T.A. Pickett and P. S. White, eds., *The Ecology of Natural Disturbance and Patch Dynamics*, Orlando, FL: Academic Press, 1985; Botkin, *Discordant Harmonies*. In extreme form, the disturbance perspective can be used politically to rationalize anthropogenic destruction of the environment; see Donald Worster, *The Wealth of Nature: Environmental History and the Ecological Imagination*, New York: Oxford University Press, 1993; Ludwig Trepl, "Holism and Reductionism in Ecology: Technical, Political and Ideological Implications," *Capitalism, Nature, Socialism* 5, 1994. But see also the response to Trepl from Richard Levens and Richard C. Lewontin, "Holism and Reductionism in Ecology," *Capitalism, Nature, Socialism* 5, 1994.

36 Michael E. Soulé and Gary Lease, eds., *Reinventing Nature? Responses to Postmodern Deconstruction*, Washington, DC: Island Press, 1995. For feminist/postmodern critiques of science, see Sandra Harding, *The Science Question in Feminism*, Ithaca, NY: Cornell University Press, 1986; Haraway, *Primate Visions*; Lynda Birke, *Feminism, Animals and Science: The Naming of the Shrew*, Buckingham: Open University Press, 1994.

37 Katherine N. Hayles, "Searching for Common Ground," in Michael E. Soulé and Gary Lease, eds., *Reinventing Nature? Responses to Postmodern Deconstruction*, Washington, DC: Island Press, 1995, pp. 47–64.

38 Daniel L. Leedy, Robert M. Maestro, and Thomas M. Franklin, *Planning for Wildlife in Cities and Suburbs*, Washington, DC: US Government Printing Office, 1978; Arthur C. Nelson, James C. Nicholas, and Lindell L. Marsh, "New Fangled Impact Fees: Both the Environment and New Development Benefit from Environmental Linkage Fees," *Planning* 58, 1992.

39 Only a small number of HCPs have been developed or are in progress, and the approach remains hotly contested. See Timothy Beatley, *Habitat Conservation Planning: Endangered Species and Urban Growth*, Austin, TX: University of Texas Press, 1994.

40 Sim Van der Ryn and Peter Calthorpe, *Sustainable Cities: A New Design Synthesis for Cities, Suburbs, and Towns*, San Francisco, CA: Sierra Club Books, 1991; Richard Stren, Rodney White, and Joseph Whitney, *Sustainable Cities: Urbanization and the Environment in International Perspective*, Boulder, CO: Westview Press, 1992; Rutherford H. Platt, Rowan A. Rowntree, and Pamela C. Muick, eds., *The Ecological City: Preserving and Restoring Urban Biodiversity*, Minneapolis, MN: University of Minnesota Press, 1994.

41 An interesting exception is the green-inspired manifesto for sustainable urban development; see Peter Berg, Beryl Magilavy, and Seth Zuckerman, eds., *A Green City Program for San Francisco Bay Area Cities and Towns*, San Francisco, CA: Planet Drum Books, 1986, pp. 48–49, which recommends riparian setback requirements to protect wildlife, review of toxic releases for their impacts on wildlife, habitat restoration, a department of natural life to work on behalf of urban wildness, citizen education, mechanisms to fund habitat maintenance, and (somewhat oxymoronically) the "creation" of "new wild places."

42 Dana Cuff, *Architecture: The Story of Practice*, Cambridge, MA: MIT Press, 1991.

43 Evernden, *Social Creation of Nature*, p. 119.

44 Richard T. T. Foreman and Michel Godron, *Landscape Ecology*, New York: John Wiley and Sons, 1986.

45 Charles E. Little, *Greenways for America*, Baltimore, MD: Johns Hopkins University Press, 1990; Daniel S. Smith and Paul Cawood Hellmund, *Ecology of Greenways: Design and Function of Linear Conservation Areas*, Minneapolis, MN: University of Minnesota Press, 1993.

46 There is also scientific debate about the merits of corridors; see, for instance, Daniel Simberloff and James Cox, "Consequences and Costs of Conservation Corridors," *Conservation Biology* 1, 1987: Simberloff and Cox argue that corridors may help spread diseases and exotics, decrease genetic variation or disrupt local adaptations and coadapted gene complexes, spread fire or other contagious catastrophes, and increase

exposure to hunters/poachers and other predators. Reed F. Noss, "Corridors in Real Landscapes: A Reply to Simberloff and Cox," *Conservation Biology* 1, 1987, however, maintains that the best argument for corridors is that the original landscape was interconnected.

47 Jennifer Wolch and Stacy Rowe, "Companions in the Park: Laurel Canyon Dog Park, Los Angeles," *Landscape* 31, 1993.

48 Such practices include putting large numbers of companion animals to death on a routine basis, selling impounded animals to biomedical laboratories, and so on.

49 Alain Touraine, *The Return of the Actor: Social Theory in Postindustrial Society*, Minneapolis, MN: University of Minnesota Press, 1988; Alberto Melucci, *Nomads of the Present: Social Movements and Individual Needs in Contemporary Society*, Philadelphia, PA: Temple University Press, 1989; Alan Scott, *Ideology and the New Social Movements*, London: Unwin Hyman, 1990.

6 Kansas City

The morphology of an American zoöpolis through film

Julie Urbanik

Popular and academic histories of cities have traditionally been told through a human-centric lens with humans portrayed as the only actors who, by their power and spirit, literally and culturally (re)create the urban. This human-centric lens has been paralleled in geography since Carl Sauer first defined the cultural landscape as the result of the process by which human culture acts as the agent upon the natural landscape (Sauer, 1925). While the cultural landscape concept itself has become more nuanced through explorations of how power, symbolism, and identity intersect with landscape, the underlying view of human exceptionalism has remained (Plumwood, 2006; Whatmore, 2006).

As animal geographers, however, we know that when we look more closely at cities we find that right next to humans are all sorts of other animals – whether as food, companions, wildlife, workers, symbols, or representations. Indeed, a city is not a metropolis but a zoöpolis (Wolch, 1995, and this volume). And animal geographers have made many powerful contributions toward understanding the human–animal-polis nexus, but according to Lestel, Brunois, and Gaunet, "we still need work that attempts to account for the shared lives that grow up between humans and animals. Simply studying the effect one has on the other is not enough" (2006, p. 156).

This chapter addresses their request by using a *zoocultural landscape* framework to map the morphology of Kansas City, Missouri, as a zoöpolis, through film. Like other cities, Kansas City is no exception to the anthropocentric historical narrative (Montgomery & Kasper, 1999), but by taking an historical animal geography approach and combining it with the narrative opportunities of digital media, there is an opportunity to expand the reach of this field and to visually re-animate urban histories. The chapter opens by providing context for urban animal and visual animal geographies, then outlines the zoocultural landscape framework before moving into the discussion of how the film *Kansas City: An American Zoöpolis* reveals the "shared lives that grow up between humans and animals" in place.

Urban animal geography – the case for moving from snapshot to narrative film

Jennifer Wolch (1995) coined the term *zoöpolis* to provide more precise language with which to reframe the urban as more-than-human. In addition, she has provided a three-part framework for helping uncover conceptual linkages between humans and animals: "how animals shape identity and subjectivity, the role of animals and urban place formation, and the evolution and dilemmas that arise when animals are allowed to figure in our urban moral reckoning" (Wolch, 2002, p. 726). Ongoing work in urban animal geography has reflected these main themes, for example, through historical analysis of zoos, pets, and livestock (Anderson, 1995; Howell, 1998; Philo, 1998) and present-day case studies of livestock and pets (Hovorka, 2008; Nast, 2006; Urbanik & Morgan, 2013).

This intersection of animal and urban geographies has fundamentally changed what we *understand* as the urban, but if we are to follow the editors of this collection's call to explore how spatially situated human–animal relations have changed through time, then historical animal geographers need to expand the currently dominant trajectory of producing textual "snapshots" of *specific* urban human–animal relations with methods that help us understand and *see* the ongoing "becomings" of a zoöpolis across *all* types of relations (Schein, 1997). In other words, if we could map both the topological (relational) and the topographical (physical contours and coordinates) of the "shared human–animal lives" of a zoöpolis through history, then we strengthen the challenge to anthropocentric narratives (both text and visually based) of cities.

Work on visual animal geographies is becoming more established (Brown & Dilley, 2012; Collard, 2016; Davies, 2000; Lorimer, 2010; Ryan, 2000; Whatmore, 2002). Methodological questions about production expertise, nonhuman animal consent, and privileging the human gaze sit alongside recognized opportunities for more deeply engaging with the relations between humans and other species. However, as with the urban animal geography work, these visual research projects have also been "snapshot" based. It is not to say that "snapshot" research is not important – indeed, it is essential, but it is more useful as a comparative opportunity between specific relations than as a constructive opportunity for deep and dwelling-in-place analysis. As historical animal geographers, this chapter posits that we must demonstrate the essentialness of nonhumans to place biographies textually *and* visually. But how to track and trace the morphologies of a zoöpolis? Where to look? What to look for? Tracking and tracing the intersecting trails of *zoocultural landscapes* provides a framework for synthesizing the shared lives.

The zoocultural landscape

Since Sauer, studies of cultural landscapes have become a central part of geographic inquiry as a whole and in the development of the second wave of animal geography (Urbanik, 2012).

> A cultural landscape is the successive alteration over time of the material habitat of a sedentary human society responding with growing strength and variety to the dynamic challenges of nature, the society's own needs and desires, and the historical circumstances of different regions in different times.
>
> (Conzen, 2001, p. 3068)

Furthermore, "The cultural landscape, as both a material presence and conceptual framing, serves to discipline interpreting subjects alongside their objectification of landscape's form and meaning" (Schein, 1997, p. 662). Put differently, cultural landscapes reflect the meanings of social life. In fact, "landscape is not only something we see, it is also a way of seeing things, a particular way of looking at and picturing the world around us. Landscapes are not just about what we see but about how we look" (Wylie, 2007, p. 7). Prefixes such as moral, political, ethnic, economic, and urban before "landscape" are now a common part of our human geographic vocabulary and help us know where, how, and why we are looking.

The difficulties of uncovering specific historical animal geographies are clear from work in this volume and in other histories of urban animals (Atkins, 2012; Biehler, 2013; Brown, 2016). But select human–animal relations in time and space is not the same as the idea of successive alterations. It follows that we need a way to "see" the full spectrum of these intersections *simultaneously* to reinforce the ways in which human–animal relations shape the meanings of social life in ways parallel to the myriad human–human relations. Following Wolch's neologism of zoöpolis to define a multispecies city, I define the *zoocultural landscape* as the compilation of the material and conceptual impresses of multispecies intersections in places.

To visualize this, we can return to Sauer's original graphic of the cultural landscape in which he saw "Culture" as the agent acting through time on the medium of natural landscapes through the forms of population, housing, livestock keeping, etc. to produce the cultural landscape (Figure 6.1). I amended his visualization in

The Anthropocentric Cultural Landscape

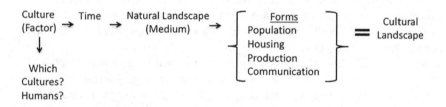

The Zoocultural Landscape

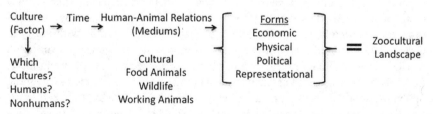

Figure 6.1 A visualization of the zoocultural landscape as a modification of Sauer's original cultural landscape methodology

Source: Sauer, 1925

the following ways: (1) by showing the expansion of the cultural landscape idea to other anthropocentric human–human relational factors within human geography; and (2) by developing the zoocultural landscape where different human–animal factors morph over time through visible forms. By excavating and rebuilding these forms, it is possible to document a more comprehensive morphology of a zoocultural landscape.

The research process was structured to categorize material into specific factors and forms as it was uncovered through primary source material in five local archives, visiting approximately 30 historical and present-day animal-related sites, reviewing secondary historical accounts, and mapping the visible human–animal daily life of the city (e.g., pet events, artwork, local media). The goal was to see if (a) enough visual material could be uncovered and (b) that a visual narrative of the city could be constructed from the disparate pieces of the zoocultural landscape framework. The answer to both was a resounding yes (with plenty of material still out there!). The next section provides an explanation of how we translated the raw material into the visual narrative that became a 30-minute documentary film.

Kansas City: the morphology of an American zoöpolis

The research revealed that while Kansas City is most well-known for its "Cowtown" history at the turn of the 20th century, there are actually four major animal-based economies which, over time, directly contributed to the identity and growth of the city. The film uses these four economic periods as its structural frame and subsumes time-specific cultural and political relations within the economic frames to present the ebb and flow of the overall zoocultural landscape. More specifically, archival footage, historical images/maps, and direct historical quotes were combined with present-day footage, interviews, and graphics, which enabled stories of individual humans and animals to meld with larger categories of relations (e.g., gender and livestock) (Figures 6.2a, 6.2b).

Part One, "The Early Years," covers the time period from 1804 to 1869. While the indigenous Kansa and Osage had lived in the area for generations and the French and Spanish had moved through by the 1700s, the story of Kansas City as a zoöpolis begins with Meriwether Lewis and William Clark's Corps of Discovery expedition from 1804 to 1806 to map the Louisiana Purchase. Their first-hand experience of the animal riches of the area led William Clark to help establish Fort Osage in 1808. As the second outpost of the Louisiana territory, its purpose was to protect the newly acquired land and to facilitate the fur trade with the local Osage. The fort held a federally mandated monopoly on the trade from 1808 to 1822. Pierre Chouteau leased a cabin at Fort Osage as part of the family's St. Louis-based American Fur Company and in 1821 sent his son, François Chouteau, up the Missouri River to found a trading post to take advantage of the end of the Fort Osage monopoly. He settled first just east of what is Kansas City today, but a flood in 1826 forced him to resettle a little to the west and to the south side of the river. This location, known then as Chouteau's Landing, is considered Kansas City's birthplace (Shortridge, 2012).

Figure 6.2a Still from the film depicting an 1855 postcard of Kansas City with an overlay of the present-day city skyline and four of the city's animal sculptures sitting on top of a sculpture known as the "Pylons," which in reality do not have animals, but abstract shapes, on top

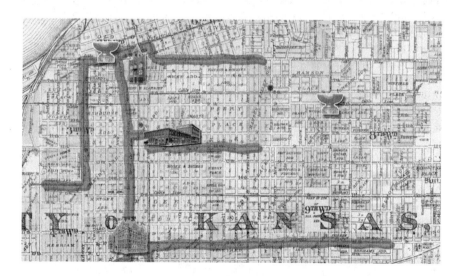

Figure 6.2b Still from the film showing the zoocultural landscape remnants on historic maps – in this case an 1890 map by G. M. Hopkins with overlays of known locations of horse fountains, horse trolley routes, the Humane Society headquarters, and fire stations. Combining the past and present gives a visual association of the historical depth of the zoocultural landscapes of the city.

Source: Permission to use historic maps and images was granted by the Missouri Valley Special Collections Room of the Kansas City Library.

From 1826 through the 1840s, Chouteau's Landing grew in population and expanded westward along the southern bank of the Missouri River toward the Kaw River where the terrain was generally flat and accessible. While this first animal-based economy in what would become Kansas City was good for European settlers, over-harvesting had a negative impact on species like the American Bison, beavers, deer, and turkey. The ultimate demise of the fur trade economy did not, however, mean the end of Chouteau's Landing. Its strategic crossroads location on a river at the western edge of the country helped it develop a second animal-based economy: that of an outfitter economy supplying the pioneers of the westward expansion along the Santa Fe and Oregon Trails with animals and goods. Settlement moved up onto the river bluffs and began tracking southwards – pushing the Indigenous peoples ever farther west. The film documents how these settlements were built not just *with* animals (mainly horses and mules), who were instrumental in the building of roads as material haulers, but *of* animals in the form of a 300-million-year-old rock layer of limestone, which was quarried and used in the foundations of residential and commercial structures.

Political and cultural connections between humans and animals during this time are noted in several ways. The first is through the chartering of the Town of Kansas in 1850. According to local legends, other names in the running were Rabbitville and Possum-trot. The second is through the role of horses during the Civil War. From October 21 to 23 in 1864, the largest Civil War battle west of the Mississippi took place about five miles south of the river. The Battle of Westport was fought by 30,000 cavalry and mounted infantry troops, with Union forces ultimately prevailing and protecting the town. Two stories of individuals provide some context for notions of welfare and human–animal bonding. One of the largest overland freighting operations was Russell, Majors, and Waddell, who managed over 1,000 men, 5,000 wagons, and 40,000 oxen. The firm was the chief supplier of the U.S. military in the West, and Alexander Majors became the first millionaire in Kansas City. Known for being an animal lover, he would not hire anyone for his overland teams that was rough with animals, but he over-extended himself with his grand idea of the Pony Express and ultimately died penniless. Another example can be found in Old Drum. Old Drum was a hunting dog who was shot, and the court case that ensued not only got reparations for the suffering of Old Drum's owner, but brought the phrase "a dog is man's best friend" into American culture.

By 1869, the Kansas City (the Town of Kansas quickly became Kansas City locally) area was well on its way. Two animal-related events mark the transition between Part One and Part Two of the film. Local businessman Joseph G. McCoy's idea in 1867 to shorten cattle drives by shipping cattle – then mainly Texas Longhorns – from Abilene, Kansas to Kansas City by rail, where they could be rested before continuing on to the packing yards in Chicago and farther east coincided perfectly with the 1869 opening of the Hannibal Bridge. It was the first to span the Missouri River and provided a direct connection between the east and the west. These developments made local businessmen realize they could bypass shipping cattle all together, and Kansas City as "Cowtown," as it came to be known, was born.

In Part Two, "The Smell of Money," the film follows the growing livestock economy with the opening of the first stockyard, the Kansas Stockyards Company, and a livestock exchange building. The stockyards grew rapidly during the next three decades, and by 1900 the modern, high-tech livestock industry was home to 14 plants and employed 35,000 people, making it the second largest meat-packing center in the nation after Chicago.

One livestock breed, the American Hereford, became so intimately connected to the identity of the city that an enormous Hereford Bull statue still sits atop the city's skyline (Sanders, 1914) in honor of the individual named "Anxiety IV." The breed itself originated in Herefordshire, England in the 18th century and was first brought to the US in 1817. It was not until 1881, when local farmers Charles Gudgell and Thomas A. Simpson imported a bull by the name of Anxiety IV that Herefords became so popular. Gudgell and Simpson revolutionized the cattle industry by breeding to take advantage of Anxiety's huge hindquarters. Previously most meat came from the front of cattle, but with his offspring most meat came from the rear, where more expensive cuts are located. Nearly all American Herefords today share Anxiety's bloodline. Gudgell founded the American Hereford Association in 1883, and not only is it one of Kansas City's longest-lived organizations, but it is responsible for the cultural icon that is the annual American Royal, the famous livestock and horse show that began in 1899 and continues today.

While the livestock economy boomed, other animals, especially horses and mules, continued to be central to the daily lives of Kansas Citians. When the Missouri Mule made its first appearance in 1904 at the American Royal, it had already been in the area for 100 years thanks to William Becknell's successful trip in 1822 that founded the Santa Fe Trail. He brought back a herd of Mexican mules, and Missouri farmers were quick to realize their potential. Mules are generally hardier and with more mellow dispositions than horses. The quality of Missouri Mules was so renowned that 350,000 of them were sent to help the British during World War I. During this time, horses were the main mode of transportation, and streets were designed to be wide enough to turn two-horse buggies around. Not only were there famous businesses like Beggs Wagon Company, F. Weber's Sons, and JJ Fosters catering to the horse economy, but there was also a world-class horse trolley system. The Kansas City & Westport Horse Railroad Company and The Jackson County Horse Railroad Company ran until 1897, when cable lines proved more profitable and less stubborn.

There was a first-class horse-based fire department under Chief George C. Hale. Two famous fire horses were Buck and Mack. They performed in 1900 at the National Fire Conference in Paris. After they won the exhibition, Chief Hale was deluged with proposals. The London Daily Mail offered $1,000 to any fire team that could equal the Kansas City team. Each day for a full week a new team from London took up the challenge, but Kansas City consistently won by performing the routine in 35 seconds. The task required harnessing the horses, running 200 yards with the engine, laying 100 feet of hose, and shooting water through a hose. The last horse team made its final emergency run in 1927.

Two important political events shaping the city over the long term also occurred during this time. In 1883, the Humane Society of Kansas City was chartered to "aid in protecting children and preventing cruelty thereto, preventing cruelty to animals, and promoting humane sentiments among all classes of persons." One of their main goals was to assist the many working horses and mules. They did this by creating the city's very first horse fountain in 1904. The film documents the known locations of these fountains throughout the city. The Humane Society also sought to protect these working animals from cruelty and abuse. Their agents declared horses unfit for work, corrected horse teams who were overloaded or otherwise abused, and made sure that horses were humanely killed when the time came. It is also at this time that several detailed anti-animal cruelty statutes are entered into the city's charter.

Kansas Citians also loved to bet on horses, and races at venues like the Interstate Fair were a popular pastime. The fair's half-mile track brought thousands of people through the gates each year from 1883 to 1886. They made "donations" and got "refunds" since gambling was illegal. It was just such a place that changed the course of the city's political history. Jim Pendergast was a betting man and won big on a horse named Climax. He used the money to buy a hotel and saloon in the West Bottoms and named the saloon after the lucky horse. The income from this business allowed him to get into politics.

When Jim died in 1911, his brother Tom won Jim's seat on the city council. Joseph Shannon, the leader of the Democrats in the ninth ward, was the main competition to the Pendergasts. At some point, Pendergast followers became known as the "goats" while Shannon's became the "rabbits." There are different interpretations as to why, with some saying it was simply a difference in the preference of kept animals (apparently, goats were popular in the first ward), and with others saying it was simple name calling (Redding, 1947). Tom Pendergast would go on to dominant Kansas City politics during the time the city became the "Paris of the Plains" until his downfall from tax evasion in 1939.

This time also had individuals famous for their connections to animals. Tom Bass was a freed slave who moved to Kansas City in the early 1890s (Riley, 1999). One of the most famous horse trainers at the time, he used methods that emphasized gentleness and harmony rather than the often violent, strong-arm training techniques commonly in use. It was his idea to develop a fundraiser to support the KC Fire Department that ultimately led to the creation of the American Royal. Loula Long Combs was the daughter of a timber magnate and a renowned animal lover, philanthropist, and show rider. She performed at the American Royal and won many blue ribbons. Her ability to move out of traditional gender roles earned her and her harness horses respect throughout the world and, at home, she used her farm to produce milk for needy families and gave generously to the Humane Society.

Two cultural events were also key to shaping the city's identity. Thomas Swope, an animal lover and supporter of the Humane Society, had donated the land for Swope Park in 1896. By 1907 the Humane Society had begun advocating for a zoo, with Swope Park being the logical choice for its location, and local businessman Barron Fradenburg argued at the time that: "Kansas City cannot be a metropolitan without a zoological garden" (Mobley & Harris, 1991, p. 89). The zoo officially opened in December of 1909 with four lions, three monkeys, a wolf, a fox, a coyote, a badger, a lynx, an eagle, and other birds. While mice were not part of the new zoo, one particular Kansas City mouse was on his way to enduring fame as the second main cultural event during this time. Walt Disney incorporated Laugh-O-Gram Films in 1922. He often slept in his studio and befriended a resident mouse he named Mortimer. This friendship later became the inspiration for the world's most-recognized fictional character, Mickey Mouse.

Part Two ends with the story of the only force that could challenge Cowtown: the Missouri River. The devastation wrought by the flood of 1951 began a slow demise of the Cowtown economy that wouldn't officially end until 1991, when the bits and pieces of the physical infrastructure that processed millions of animals, employed thousands of people, and made a few people very rich were disassembled like the animals themselves. But that obviously did not end human relationships with other animals in the city.

Part Three, "The New Animal Economies," narrates the two main legacies to Cowtown and a changed but familiar zoocultural landscape. The first is the return of livestock production. Historically, residents not only consumed livestock meat, dairy, and poultry products, but they often lived with these animals on urban farms. But post-World War II, changing cultural norms about hygiene and food safety coincided with the flood of 1951 to remove farm animals from daily city life (in a different context, see Chapter 8 on pigs in London in this volume). Over the past decade, an increasing number of producers are reviving urban and local farming because consumers are becoming more interested in where their animal products come from, how the animals are raised, and how they can support family farmers. The second Cowtown legacy represents a shift for working animals. The horses and mules of yesterday's street scenes have given way to research animals residing indoors. Since 2006, the Animal Health Corridor has brought together 300 participating businesses, educational institutions, governments, and trade groups to promote the greater Kansas City region as the largest concentration of the animal health industry in the world. The economic impact of the Health Corridor is equal to the livestock industry. Companies located within the Animal Health Corridor annually represent 82%, or $7.4 billion, of total U.S. Animal Health and Diagnostic Sales and 61%, or $14 billion, of the total pet food sold in the U.S.

The Animal Health Corridor is also part of another form of animal-based economies in the city: the pet economy. The pet economy includes the full gamut of vets, groomers, pooper-scoopers, stores, sitters, boarders, trainers, photographers, technologies, events, cemeteries, and dedicated parks. Indeed, public horse stables and fountains have given way to dog parks, dog fountains, and doggie day care centers. Several companies have their origins in the city. Three Dog Bakery, which opened in Kansas City in 1989 and is now nationwide, provides hand-crafted dog treats in all manner of shapes and sizes. Greenies are hard treats designed to help clean dogs' teeth and freshen breath. The product was launched in 1998 by locals Joe and Judy Roetheli and then bought by the Mars Corporation. Neuticles is a company that makes testicular implants for pets and livestock for people who want their animals to look "intact" after they have been neutered. Over half a million pet owners have chosen to have this type of surgery since their debut in 1995. And Fitbark is the newest startup making huge waves for its wearable activity monitors for pets.

Since the founding of the Humane Society, Kansas Citians have also deeply cared about the welfare of animal companions. Today there are many shelters in the area as well as a long list of breed/species-specific rescues. KC Pet Project, while privately operated since 2012, is the public face of the city's management of pets. They are the third largest open-admissions shelter in the country to receive no-kill status. Spay/Neuter Kansas City focuses on the low-income community and provides assistance to families who may not always have the financial means to care for their pets.

The city also continues to enjoy a variety of animal celebrities. From the original superstars of Buck and Mack, Anxiety IV, and Mortimer to the sports mascots of the KC Royals baseball team's Charlie O and Sluggerrr, and the KC Chiefs football team's Warpaint and KC Wolf, Kansas City basks in multispecies pride. In addition, there is also quite a zoo of public art sitting on those limestone foundations documenting the variety of human–animal relations. Some of the animals that have joined The Scout (Sioux Indian on horse, 1922), The Pioneer Mother (a pioneer woman and her family with horse, 1927), and BOB (Hereford Bull statue, 1951) over the years include an eagle, lions, a penguin, a bison, some spiders, two sea horses, a zebra, a boar, several nonhuman hipster murals, and Seaman, the Newfoundland dog who accompanied Lewis and Clark on their entire journey.

What of the wildlife that started it all? The last section of the film shows that while the fur-trapping industry is nowhere near as big as it was in Kansas City's early days on the river, an economy based on hunting, trapping, furs, and taxidermy still flourishes through companies such as Bass Pro, Cabela's, and Oracle, a fine art/taxidermy boutique. Politically, the city works closely with the Missouri Department of Conservation to manage wildlife. They have reintroduced the peregrine falcon to the area because people love peregrines for their beauty, their speed, and especially their love of pigeons as a food source. When it comes to deer, however, the city has the opposite problem. There are so many it is hard to believe that they were once scarce due to overhunting! Managed hunts in Swope Park remove around 500 deer per year, which reduces human–deer conflicts and also ensures that deer do not succumb to starvation or disease. And the city-owned Lakeside Nature Center is a wildlife rehabilitation hospital that also educates residents about the lives of native species.

This all-too-brief narrative summary of film hopefully provides enough information to demonstrate that Kansas City would not be Kansas City without its nonhuman histories. It is a city that has loved its animals, worked its animals, consumed its animals, and celebrated its animals. Its physical, political, cultural, and economic landscapes are deeply entwined with nonhuman relations. The supporting visuals allow the viewer to experience Kansas City as a zoöpolis by "seeing" the changing forms in the movements of animal economies from the edge of the river during the fur trade, to the inland locations of the outfitters and their farms, to the development of the livestock industry in the west bottoms, and the dispersed animal sciences and pet economies of today.

Reflections on the zoocultural landscape framework and the process of the film

This attention to the morphology of zoocultural landscapes reveals that there is a lot of information to be found and "read," and that comparing the factors of relations and their forms over time provides a methodology for going beyond the historical snapshot approach to exploring the full complexity of relations *in* a place and also *as* a place. The film was shown in a variety of local venues (including being selected for the first local film showcase for the annual Kansas City Film Festival in 2016) and is now available for free online. Feedback from viewers has been quite positive overall, with several commenting that they had lived in Kansas City all of their lives and they did not know half the information in the film. What was particularly fascinating to viewers, however, was the fact that so many different types of human–animal relations had been going on for so long. Many people thought, for example, that animal welfare in the form of shelters and current pet anti-cruelty laws was a modern-day part of the city, so to learn that there was a humane society in 1883 and that the first anti-cruelty statutes were implemented in the 1928 city ordinances was revelatory. Others had no idea there were such things as public horse fountains or that streets had been built to allow horse carriages to turn around. Finally, many commented that it was interesting to watch an "animal movie" that was completely different from the overtly political animal rights exposés and human-less nature films. The film is now also being used successfully as course material in animal geography-related classes (Doubleday, 2017).

Because this project was completed with a limited budget, there was quite a bit of material that was left undeveloped but remains an important part of building out zoocultural landscapes going forward. For example, a detailed compilation of tri-angulated information from city directories, the Sandborn Fire Insurance maps and city datasets could provide near block-by-block changes in zoocultural landscapes of the economic relations. A superficial examination of local toponyms revealed tantalizing opportunities to map in more detail street names, parks, and neighbor-hoods as they might relate to individual humans, animals, or events. Even acts of rebellious animal agency, such as escaped cows or zoo animals, could be mapped onto the city from news archives if there was enough time and eye power to scroll through microfilm. In other cases, particular stories were left out of the film because they lacked enough visual material (e.g., the early history of the rise of BBQ) to fill narrative timing.

What made this project successful as a work of historical animal geography and as a media product for the public was the ability to successfully bring forward the stories of human–animal relations while simultaneously locating them in the changing zoocultural landscapes. The zoocultural landscape framework worked well to corral the bits of material as they were uncovered. By learning where, how, and what to look for in the zoocultural landscapes of Kansas City, the opportunity to bring animals into the zoöpolis for everyone to "see" was manifest. The concept of the zoocultural landscape and the use of film-based storytelling helps begin to tell the "shared lives of humans and animals in places," as well as demonstrating where and how particular human–animal configurations fit into Wolch's framework of urban identities, urban place formation, and urban moral reckonings.

References

Anderson, K. (1995). Culture and nature at the Adelaide Zoo: At the frontiers of 'human' geography. *Transactions, Institute of British Geographers*, 20, 275–284.

Atkins, P. (2012). *Animal Cities: Beastly Urban Histories*. Surrey, England and Burlington, VT: Ashgate.

Biehler, D. (2013). *Pests in the City: Flies, Bedbugs, Cockroaches, and Rats*. Seattle: University of Washington Press.

Brown, F. (2016). *The City Is More Than Human: An Animal History of Seattle*. Seattle: University of Washington Press.

Brown, K. and R. Dilley. (2012). Ways of knowing for 'response-ability' in more-than-human encounters: The role of anticipatory knowledges in outdoor access with dogs. *Area*, 44, 37–45.

Collard, R.-C. (2016). Electric elephants and the lively/lethal energies of wildlife documentary film. *Area*, 48(4), 472–479.

Conzen, M. P. (2001). Cultural landscape. In: Smelser, N. and Baltes, P. (Eds.), *International Encyclopedia of the Social and Behavioral Sciences* (pp. 3086–3092). Oxford: Pergamon.

Davies, G. (2000). Virtual animals in electronic zoos: The changing geographies of animal capture and display. In: Philo, C. and Wilbert, C. (Eds.), *Animal Spaces, Beastly Places: New Geographies of Human-Animal Relations* (pp. 243–267). New York and London: Routledge.

Doubleday, K. (2017). Teaching animal geographies. *Animal Geography Specialty Group of the AAG Newsletter*, 8, 8. www.animalgeography.org/newsletter.html.

Hovorka, A. (2008). Transspecies urban theory: Chickens in an African city. *Cultural Geographies*, 15(1), 95–117.

Howell, P. (1998). Flush and the banditti: Dog-stealing in Victorian London. In: Philo, C. and Wilbert, C. (Eds.), *Animal Spaces, Beastly Places: New Geographies of Human-Animal Relations* (pp. 35–55). New York and London: Routledge.

Lestel, D., F. Brunois and F. Gaunet. (2006). Etho-ethnology and ethno-ethology. *Social Science Information*, 45, 155–177.

Lorimer, J. (2010). Moving image methodologies for more-than-human geographies. *Cultural Geographies*, 17, 237–258.

Mobley, J. and N. Harris. (1991). *A City within a Park: 100 Years of Parks and Boulevards in Kansas City, Missouri*. Kansas City: American Society of Landscape Architects and Kansas City, Missouri Board of Parks and Recreation Commissioners.

Montgomery, R. and S. Kasper. (1999). *Kansas City: An American Story*. Kansas City: Kansas City Star Books.

Nast, H. (2006). Puptowns and Wiggly Fields-Chicago and the racialization of petlove in the twenty-first century. In: Schein, R. (Ed.), *Race and Landscape in America* (pp. 237–251). New York: Taylor & Francis and Routledge.

Philo, C. (1998). Animals, geography, and the city: Notes on inclusions and exclusions. In: Wolch, J. and Emel, J. (Eds.), *Animal Geographies: Place, Politics, and Identity in the Nature-Culture Borderlands* (pp. 51–70). London and New York: Verso.

Plumwood, V. (2006). The concept of a cultural landscape: Nature, culture and agency of the land. *Ethics & the Environment*, 11(2), 115–150.

Redding, W. M. (1947). *Tom's Town: Kansas City and the Pendergast Legend*. Philadelphia, PA and New York, NY: J. B. Lippincott Company.

Riley, K. (1999). *Biography of Tom Bass*. Kansas City: Kansas City Public Library.

Ryan, J. (2000). 'Hunting with the camera': Photography, wildlife and colonialism in Africa. In: Philo, C. and Wilbert, C. (Eds.), *Animal Spaces, Beastly Places: New Geographies of Human-Animal Relations* (pp. 205–222). New York and London: Routledge.

Sanders, A. (1914). *The Story of the Herefords an Account of the Origin and Development of the Breed in Herefordshire, a Sketch of Its Early Introduction into the United States and Canada, and Subsequent Rise to Popularity in the Western Cattle Trade, with Sundry Notes on the Management of Breeding Herds*. Chicago: Breeder's Gazette.

Sauer, C. O. (1925). The morphology of landscape. *University of California Publications in Geography*, 2, 19–54.

Schein, R. (1997). The place of landscape: A conceptual framework for interpreting an American scene. *Annals of the Association of American Geographers*, 87(4), 660–680.

Shortridge, J. (2012). *Kansas City and How It Grew, 1822–2011*. Lawrence: University Press of Kansas.

Urbanik, J. (2012). *Placing Animals: An Introduction to the Geography of Human-Animal Relations*. Lanham, MD: Rowman & Littlefield.

Urbanik, J. (2015). *Kansas City: An American Zoöpolis*. Film. Kansas City: The Coordinates Society. www.coordinatessociety.org/kczoopolis.

Urbanik, J. and M. Morgan. (2013). A tale of tails: The place of dog parks in the urban imaginary. *Geoforum*, 44, 292–302.

Whatmore, S. (2002). *Hybrid Geographies: Natures, Cultures, Spaces*. London: Sage.

Whatmore, S. (2006). Materialist returns: Practising cultural geography in and for a more-than-human world. *Cultural Geographies*, 13(4), 600–609.

Wolch, J. (1995). Zoopolis. *Capitalism, Nature, Socialism*, 7(2), 21–47.

Wolch, J. (2002). Anima urbis. *Progress in Human Geography*, 26(6), 721–742.

Wylie, J. (2007). *Landscape*. Abingdon, Oxon: Routledge.

7   The strange case of the missing slaughterhouse geographies

Chris Philo and Ian MacLachlan

> You have just dined, and however scrupulously the slaughterhouse is concealed in the graceful distance of miles, there is complicity.
> – Ralph Waldo Emerson, "Fate" in *The Conduct of Life*, 1860

Introduction

The study of animal geographies in its new guise – as a critical study of human–animal relations, attentive to the place of animals as more than just "natural" *objects* in the world – has come a long way since the 1995 *Society and Space* theme issue that first defined the field (Wolch and Emel, 1995). The present volume is ample testament to that development. Arguably, however, a key concern for the earliest work on new animal geographies (e.g. Philo, 1995, 1998; Ufkes, 1995, 1998; Watts, 2000, 2004), informing an initially quite overt animal liberationist ethico-politics in the likes of Wolch and Emel's (1998a) *Animal Geographies* anthology, has since gone missing: namely, attention to the mass killings of nonhuman animals routinely occurring in specialist spaces set aside for this express purpose, so-called slaughterhouses, abattoirs, or meatpacking plants (see Box 7.1). A restated "attention to the violent power relations at work in human–animal encounters" (Collard and Gillespie, 2017, p. 2) informs the Gillespie and Collard (2017) *Critical Animal Geographies* collection, and here slaughterhouses – and related spaces, such as the farmed animal auction, where the dairy cow "downer" is dispatched with a gunshot to the head (Collard and Gillespie, 2017, p. 1) – do make an appearance. That said, it is only in the chapter on the Hudson Valley Foie Gras "industrial animal facility" where, "after all, the ducks are ultimately killed" (Joyce, Nevins and Schneiderman, 2017, pp. 93 and 97), that such spaces are explicitly foregrounded in quite the manner proposed in the present chapter. For reasons about which we will speculate in conclusion, what we characterize as an institutionalised geography of animal death, itself largely a hidden, covert geography "concealed in [Emerson's] graceful distance of miles," has remained on the fringes of scholarship in new animal geographies.

Consider the excellent 2012 book-length overview of animal geographies by Urbanik, a remarkably detailed introduction to the field, notably to what she terms "third wave" animal geography post-1995. When prefacing her book, Urbanik

---

Box 7.1 Relevant definitions

(a) A **slaughterhouse** is a small building equipped and used for killing and dressing out small numbers of food animals. The term is not now used by meat processing professionals; indeed, they avoid it: slaughterhouses are regarded as archaic, unsanitary, inefficient and pre-industrial. Private slaughterhouses may be euphemistically described as abattoirs.

(b) **Abattoir** – from the Old French, *abattre*, to beat down – is a public or municipal killing facility of a size appropriate to the community it serves and operated by local butchers. Abattoirs are dedicated to animal slaughter, carcass dressing (removal of limbs, head, hide and viscera) with value-added processing confined to tripe and other organ meats.

(c) In North American parlance, a **packing plant** or **packing house** is a large-scale industrial establishment in which slaughter is typically performed on a "kill floor," processing animals into meat. Packing plants go beyond carcass-dressing to break (fabricate or disassemble) the carcass into primal cuts and preserving/packaging the meat for long-distance shipment to supermarkets and food service providers.

---

invites us, the humans, to contemplate the animals that "surround" us all the time, including perhaps "a freezer full of chicken or a trash can full of hamburger wrappers" (Urbanik, 2012, p. xi). Yet, the bloody spaces of slaughter that necessarily precede the chicken in the freezer or the beef in a bun then remain strangely absent. Not entirely so, as references to humans eating or otherwise using the body-parts, skins or furs of animals do recur, alongside discussions of industrial agriculture, diverse technologies "down on the farm," human health/consumption issues and animal welfare advocacy, where animal slaughter is consistently implicated if not tackled foursquare. Urbanik (2012, p. 125) acknowledges that "[t]he predominant process that puts them [animals] on our plates . . . is not very appetizing upon closer inspection," before mentioning the fate of chickens and beef cattle in crowded battery cages or feedlots prior to slaughter. Brief mention is made of Upton Sinclair's 1906 novel, *The Jungle*, which "depicted the lives and conditions of Chicago's meatpacking district[,] revealing the horrific conditions of the workers and animals" (Urbanik, 2012, p. 104), while Philo's (1998) argument is repeated about the nineteenth-century removal of livestock animals from the city "because they were no longer seen as being in the right place" (Urbanik, 2012, pp. 108–109; also Blue and Alexander, 2017, p. 155; Giraud, 2017, p. 40). The significance of slaughterhouses to Philo (1998) is rather left hanging, however, while the visceral geographies of meatpacking evoked by Sinclair are unelaborated. Although lacking space here to illustrate this claim in detail, our conviction is that – as illustrated by Urbanik (2012) – the field tends to prefer pets in domestic spaces, feral animals in the city, livestock *living* "in the fields," wild animals in the countryside and even animals in zoos and laboratories. The deathly sites of animal termination, what Wolch and Emel (1998b, p. xi) call the "death camps," remain distinctly "unappetizing."

There is work undertaken *explicitly* on slaughterhouse geographies, but it has comprised a sub-field of inquiry proceeding parallel to – rarely intersecting with – the mainstream of new animal geographies as betokened by Urbanik (2012). Indeed, there are studies, papers and authors in this sub-field almost entirely absent from the citations in new animal geographies, and the subfield has more of a presence in relation to what might be broadly construed as the sub-disciplines of economic geography (addressing the industrial and labour geographies of the meatpacking industry, mainly in a North American context: e.g. Brody, 1964; Curran, 2001; Fields, 2003, 2004; Furuseth, 1997; Page, 1998; Walsh, 1978), rural geography (addressing the role of slaughterhouses within the patterning of rural settlement and land use: e.g. Broadway, 2000; Broadway and Stull, 2006; Stull and Broadway, 2012) and food geography (addressing the origins of the animal products comprising many different foodstuffs: e.g. Buller and Morris, 2003; Buller and Roe, 2014, 2018; Miele and Evans, 2010; Roe, 2010, 2013, 2016). Tellingly perhaps, a substantial collection edited by a geographer (Atkins, 2012a) and containing impressive contributions by historical geographers (Atkins, 2012b, 2012c, esp. pp. 82–90; Laxton, 2012), one where slaughterhouses, "animal nuisances" and "animal waste" are never far from the surface, is chiefly framed as a new departure for the study of urban history, not geography.

We cannot encapsulate this whole sub-field, but in what follows we bring news of certain core work on the locational dynamics of slaughterhouses back to new animal geographies, while offering an outline of a historical geography of slaughterhouses, chiefly in Europe and North America (cf. Lakshmi Singh et al., 2014).[1] This survey emphasises the gradual (and ongoing) concealment of these spaces, buttressed by the complicity of human silences around these spaces, and we draw particularly upon historical and geographical work by one of the present co-authors, Ian MacLachlan.[2]

Putting the fleshers at a distance and/or behind high walls

The earliest urban slaughterhouses were small buildings, often close to residences and sometimes located in the butcher's own back parlour (Ayling, 1908, p. 53). Complaints about the social impact of animal slaughter frequently led to the forced relocation of slaughterhouses from inner-city residential neighbourhoods to the urban margin, close to the open countryside. In the medieval period, animal slaughter was occasionally proscribed within the walls of the city, forcing livestock processing, markets and pens to the area immediately outside the city gates. Once cities began to grow in the modern period, the urban fabric quickly expanded beyond the walls, encircling slaughterhouses and cattle markets. Land use conflicts and complaints about the presence of farm animals or the by-products of slaughter and carcass-breaking eventually prompted the relocation of the slaughterhouse beyond even these expanded urban margins. Many cities made famous by live-stock markets and animal slaughter have been through one or more cycles of this process; examples include London, Paris and Chicago (Forshaw and Bergström, 1990; Claflin, 2008). The perishability of meat nonetheless posed a locational challenge in the pre-industrial era before mechanical refrigeration, for slaughter and carcass-breaking still had to be sited close enough to urban consumers to avoid decomposition and devaluation of the meat product, which was especially impor-tant in summer months and at low latitudes.

Hence, there was a limit to the number of "graceful" miles that could be put between slaughterhouses and the city's human populations. Nonetheless, nineteenth-century slaughterhouses of all types came under pressure to relocate: "The abattoir was banished to the perimeter of the city: 'on the outskirts of the town, although the work of slaughtering, if properly carried out, should not be a nuisance to the neighbours'" (Otter, 2008a, p. 96 with a quotation from Ayling, 1908). A useful summary is given by Blecha (2015, p. 35):

> the historical geography of slaughter has . . . received recent attention. Schol-ars of urban geography and history have demonstrated how, in many cities, independent butchers (sometime organized as a guild) were removed from city centres to their edge; this move was often accompanied by shifts in organ-isation and increases in scale and mechanisation. Commercial slaughterhouses appeared at the urban fringe where the traditional butchers' craft was replaced by the (dis)assembly line.

This quote anticipates several key themes for what follows, notably a long-term locational logic pushing slaughterhouses away from the city centre into suburban or even rural settings, as suggested by the background landscape of Figure 7.1, coupled with reducing numbers of facilities and a growing scale of activities on fewer sites. Specially, for Blecha, it also serves to contextualise her own research on a more recent *return* of slaughter to the city in the guise of small-scale, do-it-yourself "backyard slaughter," which some urban-dwellers believe is a "right," but "others find . . . abhorrent and want it banned, or at the very least kept out of resi-dential neighbourhoods" (Blecha, 2015, p. 34; also Blecha, 2007; Blecha and Davis, 2014; Blecha and Leitner, 2014).

Paralleling Blecha's account, MacLachlan (2005) identifies shifting historical regimes in the spatial organisation of animal slaughter in Edinburgh, Scotland. The earliest records note the presence of the Incorporation of Fleshers, a guild dating to the fifteenth century, with slaughter taking place in or behind retail butcher shops in the "Lawnmarket" (still at the heart of the city). By the early seventeenth century, the effluvium of slaughter was acknowledged as a nuisance, and from 1622 Edin-burgh's Fleshers were proscribed from operating slaughterhouses within the city and from discharging blood or offal into the streets. Individual slaughterhouses were relocated along the shore of the North Loch, now the railway right-of-way, close to the city centre yet outside its bounds. By the late eighteenth century,

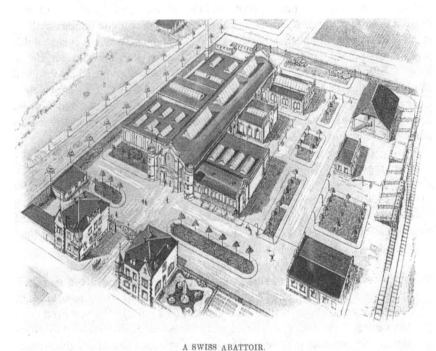

A SWISS ABATTOIR.

Fine abattoir at La Chaux de Fonds, Switzerland, mentioned by Herr Heiss.

Figure 7.1 View of the late nineteenth-century abattoir in La Chaux de Fonds, Switzerland.
Note the suggestion of rural surrounds and also its integrated circulation plan,
with vaulted concrete roadways running between and through the separate
buildings and direct railway access, the latter for incoming animals, which
hence were rendered invisible to surrounding populations even at the moment
of entering the facility.

Source: Heiss, H. (1907) The German Abattoir. In C. Cash (Ed.) (1907) *Our slaughter-house system:
a plea for reform* (p. 149). London: George Bell.

Edinburgh was growing rapidly, the construction of New Town, Edinburgh's bold
experiment in Georgian suburban design, was well underway, and what had been
an eccentric location on the city's northern margin now became its centroid. Worse,
the valley of the North Loch was spanned by a massive new viaduct, giving a bird's-
eye view of the slaughterhouses. Once again, the sights and smells of slaughter were
offending the sensibilities of urban residents. According to a 1784 pamphlet: "The
slaughter houses, in their present situation, are justly considered as the greatest and
most offensive nuisance that ever disgraced the capital of a kingdom" (quoted in
MacLachlan, 2005, p. 57). To pre-empt further complaint and avoid imposed relocation, the Fleshers took the initiative to erect a two-storey shambles.

Building designs called for twenty-feet-high walls of dressed stone,

so the business of slaughtering may not give offence to persons going or coming along the bridge to or from New Town. . . . [I]t will occasion a

considerable degree of trouble even to a curious person to be satisfied of what is going on within the walls.

(quoted in MacLachlan, 2005, p. 62)

The city was at liberty to grow around and engulf the slaughterhouses, robbing them of their privacy and seclusion, but the Fleshers managed to alter the built environment and veil the process that was their craft, still operating in the midst of the city, close to the present site of Waverly Station's main concourse. Edinburgh's first shambles was among the earliest butcher-owned abattoirs in Britain, and they were owned and operated by the Fleshers until they yielded to the inevitability of the Edinburgh and Glasgow Railway in 1846. For six years after 1846, Edinburgh regressed from having an efficient, central, purpose-built shambles to an unregulated collection of 78 separate killing booths operated by about 150 butchers dispersed throughout the city. Eventually, though, processing of food animals converged on a single new municipal abattoir, the Fountainbridge Slaughterhouse, officially opened in 1853. Though formally named a "slaughterhouse," it was truly an abattoir:

> It was a state-of-the-art facility with 28 foot walls to conceal any evidence of slaughter from the street. Far in excess of what was necessary to contain livestock or to block the view from the street, the massive walls demonstrate the lengths to which Victorians would go to conceal the act of slaughter and prevent nearby residents from exposure to the sight of fresh-dressed carcases. . . . The Fleshers were able to prosecute their urban trade in the heart of the city with little intervention provided that they concealed their carcase processing activities, contained the slaughterhouse nuisance, and kept the slaughter of food animals out of sight and out of mind behind stone walls.
>
> (MacLachlan, 2005, p. 69)

The same logic of concealment was clearly operating as elsewhere, although the extent of displacement away from the inner city was here more muted, high walls substituting for the "graceful" miles.

New model facilities, slaughterhouse reform and clean, open spaces

In Europe, the public abattoir was a municipal institution established by the butchers' guilds traceable back to the Middle Ages. The Augsburg abattoir dates to 1276 and public abattoirs were a common facility in most German towns and cities by the fourteenth century (Heiss, 1907, pp. 85–86). Spain's Laws of the Indies provided for slaughterhouses and regulated their sanitation in colonial town plans by 1573 (Mundigo and Crouch, 1977, p. 255). By the eighteenth century, public abattoirs, often with quite impressive architectural features and with state-of-the-art equipment, had become ubiquitous features of the Western European townscape, most notably in Germany (Ayling, 1908, p. 70; Cash, 1907, p. 58).

In the early nineteenth century, a Napoleonic system of public abattoirs was planned for the suburban precincts of Paris, and, soon after, for every French city. Butchers were required to slaughter and dress their livestock in these public abattoirs under government supervision, while animal by-products were retained by the abattoir to offset the cost of slaughter, a practice persisting to the present day in custom slaughter plants (Heiss, 1907, pp. 85–87; Giedion, 1948; Schwarz, 1901, p. 8). Prominent here was the "separate stall" system comprising private chambers where killing and dressing took place out of the public eye, operated by butcher-craftsmen in a secluded, quiet and even intimate atmosphere. Each animal, housed in its own stall, was killed and dressed *in situ* by a single butcher without the aid of conveyors or a division of labour, a "curious symbiosis of handicraft with centralisation" (Giedion, 1948, p. 211). By the late nineteenth century, such an arrangement was being castigated as "an agglomeration of juxtaposed *tueries*," which conserved the "repulsive" habits of an artisanal tradition that went back several centuries (quoted in Lee, 2008, p. 62).

By the 1860s, this older Parisian system of small suburban abattoirs became obsolete as Paris grew beyond its walls, and a new centralized public abattoir was conceived by George Haussman, master planner of Paris. Served by railway sidings and a canal, Abattoir La Villette was a massive structure of steel and glass opened in 1867, just in time for the International Exhibition: "It became *the* abattoir, a prototype for the rest of the century, just as the boulevards and public parks of Haussman's Paris became models from which every growing metropolis of the continent took pattern" (Giedion, 1948, p. 210; see also Claflin, 2008). Characteristic was the "open hall" system based on an airy, open and naturally lighted space with massive skylights reminiscent of the Crystal Palace of 1851 (see Figure 7.2). The open atrium separated animals by species and was adjoined by lairages to accommodate animals awaiting processing, a slaughter hall and a cooling hall to preserve the finished product. Writing of La Mouche, the modernist Lyonnaise abattoir established in 1908, Lee observes: "More than any other slaughterhouse built during this period, the complex at Lyon focused the constellation of political and social concerns shaping it into a modern 'factory' incarnation" (Lee, 2008, p. 63). Industrial design played a key role in the European abattoirs of the nineteenth century, with a spatial layout engineered to be sanitary, spacious and efficient (see Figure 7.3), boasting innovative mechanisms for immobilising, stunning, suspending, bleeding, cutting, weighing and, by the 1880s, chilling the carcass (Otter, 2008a, pp. 95–97). French adherents of *abattage* at La Villette still, it seems, had no wish to emulate the Chicago model, arguably even more aggressively "modern," of large-scale livestock marketing and industrial slaughter (Claflin, 2008, p. 34).

By the late nineteenth century, a slaughterhouse reform movement gained momentum in Europe and North America. It was driven by four inter-related concerns: (1) the need for government-sanctioned meat inspection, prompted mainly by the threat of spreading bovine tuberculosis to humans; (2) the sanitary nuisance posed by the transportation and disposal of slaughter by-products in urban areas; (3) the hazard posed by livestock to pedestrians in city streets; and (4) the potential for inhumane slaughter of food animals to repress the humane sense of compassion on the part of both slaughtermen and anyone attracted as onlookers, especially small boys: "To avoid the brutalisation of children drawn to the slaughterhouse by

FIG. 24.—LA VILLETTE, PARIS. INTERIOR OF WAITING COURT.

Figure 7.2 The waiting court of La Villette, the great central abattoir of Paris built in 1867 and operated until 1974. Note the trough to keep cattle well-watered and the provision of ample natural light through the iron-framed clerestory. The manure barrow in the far distance is a reminder of the labor-intensive nature of ante-mortem animal husbandry and the need to dispose of manure over quite long distances from urban abattoirs.

Source: R. S. Ayling (1908) *Public abattoirs: their planning, design and equipment.* London, Figure 24, pp. 70–71.

curiosity, and in the interest of public morality, it was necessary to conceal these activities from the public gaze" (MacLachlan, 2008, p. 110).[3]

In an era much preoccupied with utopian designs, the Model Abattoir Society crusaded to publicise the evils of small-scale, privately owned slaughterhouses, advocating instead a system of hygienic and publicly regulated abattoirs. The Society's concern with humane slaughter was manifest in its then-radical proposal to render animals insensible *before* they entered the slaughter court, sparing the animals still alive from the vision of slaughter, dismemberment and evisceration that was to be their fate (Otter, 2008a, p. 97; MacLachlan, 2008, p. 110).

In 1901, Britain's Royal Commission on Tuberculosis recommended the closure of private slaughterhouses when public abattoirs were available; and by 1908, 136 public abattoirs were established in locations scattered throughout Britain (Ayling, 1908; Perren, 1978, pp. 91 and 155). American and Canadian proposals to build public or municipal abattoirs coincided, somewhat ironically, with the completion of large-scale corporate slaughter in Chicago's Union Stockyard district, to be considered shortly. The very fact that they were being called "abattoirs" suggests

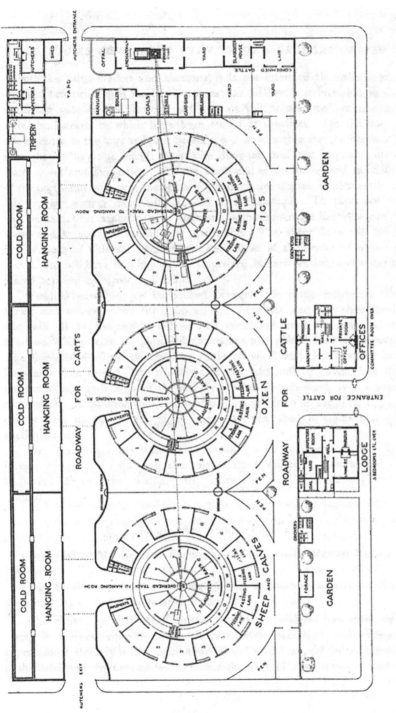

FIG. 20. PLAN BY MODEL ABATTOIR SOCIETY

Figure 7.3 A product of Benjamin Ward Richardson's London-based Model Abattoir Society, this concept plan is not to scale and was never actually built. It appears to be inspired by Jeremy Bentham's Panopticon. The three central "inspection houses" were to be staffed by a veterinary inspector able to observe and police all butchers simultaneously, ensuring that they follow the principles of humane slaughter and that there is no evidence of disease such as bovine tuberculosis.

Source: Ayling. R. S. (1908) *Public abattoirs: their planning, design and equipment.* London, Figure 20, pp. 66–67.

that the Napoleonic abattoir system held some influence, clarifying that the institution of publicly financed slaughter would share nothing with the private slaughterhouse, about which scandalous accounts abounded of unsanitary conditions and adulteration, and nothing with the "hideous multistorey mills" and mass disassembly-line production of the Chicago-style packing plants: "The American meat factory is in no sense an abattoir. The abattoir has, indeed, for its essential object the prevention of those very abuses which have made the American meat-factory a byword in the civilized world" (Cash, 1907, p. ix; see also Claflin, 2008). In the United States, public abattoirs were established in some cities, and the merits of Philadelphia's abattoir and the Brighton abattoir (northwest of Boston) were extolled in considerable detail (James, 1880; Smith and Bridges, 1982). Urban residents saw such public abattoirs as the solution to the lingering public nuisance posed by small-scale slaughterhouses, especially in residential areas, yet still keeping slaughter out of the hands of the corporate meatpacking oligopoly (MacLachlan, 2001, p. 135). Under municipal ownership, inspection and control, public abattoirs answered urban concerns about the mud, the blood and the gore of butchery, the odours, the vermin and the unseemly sights of food animals performing their natural sexual and metabolic activities in city streets (also Philo, 1995, 1998).

Urbanisation, disassembly lines and the secretive "kill floor"

The early nineteenth-century geography of United States slaughterhouses entailed a widespread dispersion of slaughterhouses catering to local demand, with almost every decent-sized town having its own slaughterhouse (Fields, 2003, p. 605). Meat-packing industrialized first in nineteenth-century Cincinnati, with both hogs and packed pork transported up and down the Ohio River, but the advent of the railroad in the late 1840s/1850s enabled the expansion of the agricultural hinterland so that the market area furnishing animals for slaughter could extend further beyond the vicinity of river ports. Rail terminals, and the growing cities which contained them, became key processing and distribution centres (Fields, 2003; Walsh, 1978), most notably and notoriously with the opening of Chicago's Union Stockyards in 1867 (Cronon, 1992). At an intra-urban scale, there were already forces afoot prompting a push *away* from city centres, while at a larger, regional scale there was a pronounced movement of slaughter *to* the largest urban centres with good transport connections.

Indeed, technological development of the North American "kill floor" here and elsewhere was dependent on high throughput based on an infrastructural and organisational triad: (1) railway transportation of livestock in cattle cars together with track-side water, feeding and confinement facilities to procure raw material from a vast hinterland and to sustain a living cargo; (2) rail-based carcass disassembly, staffed by semi-skilled workers organised into a minute division of labour; and (3) railway transportation to distribute chilled meat in reefer cars together with track-side icing stations to preserve a perishable finished product. The railway networks, cattle cars for livestock, and ice-cooled reefer cars were crucial ingredients for the large-scale disassembly of cattle and hogs. By developing and refining the railway refrigerator car, George Hammond and Gustavus Swift established the feasibility of large-scale centralized production of meat in Chicago to serve the New York market, 800 miles to the east (MacLachlan, 1998a, 1998b).

Thus, Fordist mass production was born, and the slaughter and disassembly of pigs and cattle became the inspiration for the most celebrated innovation of the twentieth century: the moving assembly line. Ford credited the overhead trolley used in dressing beef carcasses in Chicago as the inspiration for this moving assembly line, applied in 1913 to the assembly of the Model T (Ford and Crowther, 1922, p. 81). While the minute division of labour in Fordist assembly process is a common trope of industrialism, few have witnessed industrial-scale carcass breaking, hidden away in the slaughterhouse, nor is it taken as a common exemplar of Taylorism. Perhaps it was considered too "gruesome" for public exhibition in the grainy black-and-white industrial films of the early twentieth century and "not suitable for all audiences," another reason for hiding slaughter and carcass disassembly from the public gaze.

Detailed work in historical slaughterhouse geographies necessarily confronts the visceral details of the disassembly line (see Figures 7.4 and 7.5) as epitomized in the Chicago stockyards, from the "kill floor" itself where the animal

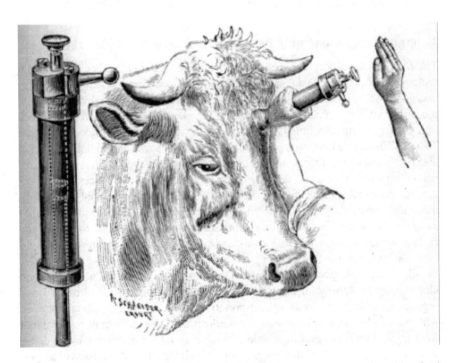

Figure 7.4 The "Flessa" shooting bolt apparatus was one of a number of cartidge-fired captive bolt stunning tools that were developed in the 1890s and the first years of the next century. Captive bolts were an advance over conventional firearms for stunning livestock because there was less chance of collateral damage to slaughterhouse workers from an errant bullet and less damage to brain tissue.

Source: Heiss, H. (1904) The stunning of cattle by means of appliances acting with instantaneous effect, translated by C. Cash (1904) *Humane slaughtering*. Coventry: Curtis and Beamish, p. 69.

ORDINARY METHODS OF SLAUGHTER.
Stunning cattle at Chicago.

Figure 7.5 Until the 1950s, most cattle were still stunned before slaughter by skilled man-
ual knockers using a pole-axe. The cattle were restrained and contained in a
rustic "knocking box" in this Chicago scene.

Source: C. Cash (Ed.) (1907) *Our slaughter-house system: a plea for reform.* London: George Bell,
p. 45.

exterminations occur, through the process of "carcass dressing" – removing the
head and hide, hair or pelt; eviscerating the body cavity – and then, once the carcass is chilled, its final breakdown into primal meat cuts (for details, see MacLachlan, 2001, pp. 140–141). At the carcass-breaking stage, the geography of the
animal itself came into play:

> The animal has been surveyed and laid off like a map; and the men have been
> classified in over thirty specialties and twenty rates of pay, from 16 cents to 50
> cents an hour. The 50 cent man is restricted to using the knife on the most delicate
> parts of the hide (floorsman) or to use the axe in splitting the backbone (splitter);
> and, wherever a less skilled man can be slipped in at 18 cents, 18½ cents, 20
> cents, 21 cents, 22½ cents, 24 cents, 25 cents, and so on, a place is made for him,
> and an occupation mapped out. . . . A 20-cent man pulls off the tail, a 22½-cent
> man pounds off another part where the hide separates readily, and the knife of
> the 40-cent man cuts a different texture and has a different "feel" from that of the
> 50-cent man. Skill has been specialized to fit the anatomy.
>
> (Commons, 1904, pp. 3–4)

This is a classic statement of the Marxian alienation of labour under industrial capitalism, the essential humanity of workers disappearing when reduced to a monetary value: "a 20-cent man." In some senses, then, it is the benighted worker who disappears, not the slaughterhouse.

After World War II, new methods and machinery were introduced to carcass dressing and meat processing (Ruttan, 1954). Mechanized hide-removing machines, power hoists and pneumatic tools such as rotary knives, power shears and band saws to split the carcass into halves contributed to technological change and labour productivity (Miller, 1956; Warren, 2007, p. 24). Then there was the development of the Can-Pak system, an on-the-rail beef dressing technology that mechanized the flow of cattle bodies through the disassembly process. It was scalable too: production could be doubled from fifteen to thirty cattle per hour simply by doubling the labour and work platforms and running the line faster; no additional floor area was required. The Can-Pak system reduced the time elapsed from the knocking box to the cooler from 50 minutes to 30 minutes (MacLachlan, 2001, p. 172). Yet, one outcome was, in effect, to start producing *too much* meat: as demand for meat lagged behind productivity growth, meatpacking employment began to fall, and post-war technological changes became the impetus for the *de*centralization of the meatpacking industry, first in the United States and later in Canada.

By 1960 most of Chicago's major multi-species, multi-storey packinghouses were silent; its stock yards finally closed down in 1970; and it might even be said that the slaughterhouse geography had reverted to the more evenly distributed, indeed ruralized, patterning of rather earlier times (Broadway and Ward, 1990; Drabenstott, Henry and Mitchell, 1999). By the 1980s, the meatpacking industry and meat-cutting employment was restructured and transformed, with the industry now being critical to the small agricultural service centres and regional economies of North America's Great Plains: places such as Brooks, Alberta; Dakota City, Nebraska; Garden City, Kansas (also Broadway and Stull, 2006); Greeley, Colorado; or Plainsview, Texas. Employing thousands of workers (many being recent immigrants) in massive kill plants visible for miles across Great Plains landscapes, the meatpacking plants of recent decades have a localized economic clout and troubled profile among social service providers that is arguably more conspicuous than it is concealed (see, for example, Stull, Broadway and Griffith, 1995; Fink, 1998; also on the relative openness of a similar facility, see Joyce, Nevins and Schneiderman, 2017).

Conspicuity of slaughter?

The conspicuity of some massive slaughterhouses on today's landscapes does offer a twist on the prevailing narrative of our chapter, wherein we have considered slaughter to be banished to the invisible margins of urban space or concealed behind walls, out of sight and mind as one of the unspeakable, unthinkable realities of our human existence. There are, however, circumstances in which slaughter is deliberately made visible and displayed prominently. As one example, the ritual slaughter of sheep, goats, camels and cattle during the *Hajj* and the *Id al-Adha* is a significant event for Muslim adherents. Animal slaughter is proscribed anywhere *other* than the five slaughterhouses of Mina, located immediately outside the "Holy Environs" of Mecca, whose daily slaughter capacity of 750,000 head is three orders of magnitude greater than for the largest industrial-scale slaughter plants in the Great Plains of the United States. Health and sanitary concerns justify slaughter plant locations of anywhere from 100 meters to 1.5 kilometres distant from the closest zones set aside for pilgrims to camp and assemble during the *Hajj*, but at the time set aside for the ritual slaughter, enormous crowds numbering in the hundreds of thousands gather to select, purchase and slaughter their animals. Roughly 50 animals are slaughtered for every 100 pilgrims; and, while some meat is consumed immediately and more is frozen for later consumption, thousands of carcasses are incinerated or buried in massive pits following their ritual slaughter (Brooke, 1987).

A second reason for the deliberate exhibition of slaughter arises from the need for regulation of slaughterhouse activities to meet concerns about humane slaughter, reduction of the public nuisance and meat inspection (Collins and Huey, 2015). Inspection was fundamental as a regulatory tool in the emergence and governance of the nineteenth-century British state (Otter, 2008b, p. 100; see also Trabsky, 2015, p. 179), but an effective inspection system was hard to operate where dispersed small-scale slaughterhouses endured: (1) because government meat inspectors could not possibly monitor multitudinous private facilities; and (2) because the sickest animals least likely to pass inspection would be diverted from inspected public facilities to more shadowy private facilities or might be farm-killed (MacLachlan, 2007, p. 246). Thus, as Otter (2006, p. 528) argues: "Slaughter had to be made public, dragged from the shadows and made visible to trained inspectors able to detect tubercles, lesions or evidence of subterfuge." The rise of a humane slaughter movement has been part of this story, seeking to expose the abuses of the secretive slaughterhouse – not just on the "kill floor," but also in the spaces occupied by animals during grain finishing for slaughter and shipment to the plant – as an input to arguments for reformed regulations, practices and indeed spatial arrangements governing the care and transportation of food animals (Buller and Roe, 2012; Johnston, 2013).[4] In a minor register, the purpose of the present chapter could be framed as enhancing the conspicuity of slaughterhouses – or, at least, of a small sub-field of geographically attuned studies concerned with and about slaughterhouses – with the goal of encouraging reflection, critique and maybe action with respect to these animal "death camps" (also White, 2017, esp. pp. 22–23).[5]

Before concluding, it is worth remarking on the experience of those humans for whom the slaughterhouse is *most* conspicuous: those working on or around the "kill floor." There is the old "Slaughterman's Creed":

Thine is the task of blood.
Discharge thy task with mercy.
Let thy victim feel no pain.
Let sudden blow bring death;
Such death as thou thyself wouldst ask.

(American Humane Association, 1908, p. 63)

Box 7.2 When slaughter goes wrong

On one occasion, Ian MacLachlan was doing a time-and-motion study on a small provincially inspected "kill floor" associated with a small-town butcher shop in southern Alberta, which is locally famous for its pork sausage. They usually kill on Wednesday mornings: thirteen hogs before coffee break and seven cattle by lunch-hour, with a workforce of three plus an inspector on the day he was present. The slaughtermen did not always display a level of professional competence congruent with their creed. His field notes record the following observation on 24 May 1995: "six bullets fired from a single-shot .22 calibre rifle before a steer was finally stunned into insensibility" (MacLachlan, 2004, p. 63).

This creed enumerates the duties of those who undertake the task of blood, an occupation practised by a social class who share something of the pariah status familiar to those dealing with funeral rites and the "offensive trades" engaged in the value-added processing of animal parts such as bone boilers, tripe boilers, tallow melters, fell mongers, knackers or leather tanners (MacLachlan, 2007, p. 230). The *Burakmin* community of Japan and the *Dalits* of India come immediately to mind, but so too the skilled butcher-craftsmen of the Medieval guilds and the *shochetim* trained, examined and certified in the ritual slaughter of food animals as specified in the *kashrut*, the Jewish dietary laws codified in the *Talmud*. Animal slaughter is thus a "mystery" shrouded in the veil of secrecy long associated with the craft. Emblematic of this mystery is the capacity to "let sudden blow bring death" in as painless, instantaneous a manner as technology and skill will allow. Yet, no matter how merciful the slaughterer, how much skill and dedication is brought to the task or how expert the training, the errant poleaxe or misplaced pneumatic stunner are inevitable hazards. The sounds of animal terror – of squealing pigs, bellowing cattle or bleating sheep and goats – tell anyone in earshot of the shameful travesty of slaughter gone sadly wrong (see Box 7.2; also Roe, 2010). Perhaps it is little wonder, then, that animal slaughter has been beyond the pale through much of the modern era, practised in private, shielded from the public gaze (Fitzgerald, 2010, p. 60), rendered *in*conspicuous.

Conclusion

In his final major work, *Negative Dialectics*, Adorno, the German critical theorist, declared the importance of "staying with" Auschwitz, with the unfathomable horrors of the Nazi "death camps" of the World War II, as a constant reminder never to let the same things happen again (Adorno, 1973). Moreover, in suggesting the need to retain a "micrological" alertness to the most intimate deeds and events integral to the experience of the "death camps," Adorno (1973, p. 366; see also Philo, 2017) added a poignant image of the scholar as akin to the child fascinated by smells issuing from "the flayer's zone, from carcasses," "for the sake of [evoking] that which the stench of the cadavers expresses." In eliding camp and slaughterhouse as, if not exactly equivalent then intensely resonating, spaces of physical and moral decay, he established a conjunction of some moment as we move now to conclude our chapter.

Central to the political theory of the Italian philosopher Agamben (1998, 1999, 2005) is the proposal that "the camp," meaning Auschwitz and various other "spaces of exception" where the usual laws preventing human abuse of other humans get suspended, operates upon and recursively produces *bare life*. The latter designates an essentially political category of humanity reduced to a state of animality: humans stripped of any political qualification as members of the *polis*, as citizens of city, nation or state, and regarded instead as barely alive, a form of life indistinguishable from the brute animals of nature (cf. Hobson, 2007). This Agambendian structuring of human–political existence, supposedly fundamental to Western thought-and-action, is highly spatialized: mapping properly political humans onto the public spaces of the city, and politically disqualified humans onto either the countryside (spaces of banishment) or "the camp" (spaces of confinement and, quite possibly, termination). Seen through these lenses, the slaughterhouse and "the camp" are profoundly intertwined, both being institutionally created, bounded and shrouded sites of slaughter for those beings whose beingness – whose very existence – really does not matter.[6] (Hence too the occasional references to animal "death camps" [also Wolch and Emel, 1998b, p. xi] seeded throughout this chapter.)

This theoretical manoeuvre aligns the study of slaughterhouse geographies with an admittedly bleak, pessimistic intellectual tradition, indexed by Adorno's "negative dialectics" and Agamben's reworking of Foucauldian "biopolitics." The latter, emphasising the governance of populations, their living or dying and the spaces in which such processes are enacted (Foucault, 2003, 2007, 2008), can be given a more hopeful, optimistic and even positive inflection – as in Lorimer's (2015) biopolitical investigations of wildlife in the Anthropocene. Indeed, just such an ethos of inquiry, propelled by a richly vitalist imperative enchanted by the ever-emergent possibilities for novel human–animal configurations, has arguably permeated much that has passed as work in animal geographies since the early 2000s. Entirely understandably, in thirsting to encounter the beastly worlds of animal liveliness in their own place-making or when transgressing human spatial orderings, the animal geographers involved here have tended to avoid the sights and sites of institutionalized animal execution (cf. some contributions to the Gillespie and Collard [2017] collection). The marginalisation of slaughterhouses to ex-urban obscurity, their limited presence in human discoursing and the relative neglect of slaughterhouse geographies as a sub-field of study – even by animal geographers – thus go hand-in-glove, all driven by a wish to remain on the side of life, hope and futures (not that of death, despair and endings). Yet, in the spirit of Adorno asking us to remain with "the flayer's zone," to take seriously what it reveals about the magnitude and detail of humanity's grotesque inhumanities, our final call in this chapter is simply for animal geographies to resist these sirens and to revisit the slaughterhouse.

Notes

1 For a parallel survey of work in the history/anthropology of slaughterhouses, see the account provided in Fitzgerald (2010).
2 The idea for this chapter was Chris Philo's, with a particular objective being to bring Ian MacLachlan's work into the fold of new animal geographies. Following discussions with MacLachlan, it was agreed that it made sense for him to become a full co-author of the chapter, and the profiling of his own work here reflects Philo's prompting.
3 Such fears have been suggested as not without basis in more recent studies: e.g. Racine Jacques (2015).
4 Conspicuity in this context also sees some slaughterhouse districts being reinvented as sites of urban heritage, recreation, tourism and entertainment. In many American cities, the areas that once were pariah districts have been planned and revitalized as urban attractions and centres of consumption, entertainment and recreation: e.g. New York's Meat Packing District (now a contender for the most glamorous neighbourhood in Manhattan, www.nycgo.com/articles/must-see-meatpacking-district-slideshow); Fort Worth Stockyards National Historic District (complete with twice-daily cattle drives); or Parc de la Villette (site of Europe's largest science museum, redesigned and repurposed based on suggestions by Jacques Derrida).
5 "If slaughterhouses had glass walls, we'd all be vegetarians" (Rasmussen, 2017, p. 54). This quote – attributed to various sources – is used by Rasmussen (2017) in a provocative piece exploring the strange entangling of so-called ag-gag laws, US prohibitions against the likes of animal activists filming/photographing cruelty against animals in industrial animal facilities, and laws against making "crush videos," pornographic films depicting sexual acts involving animals.
6 We did not see the excellent forthcoming monograph by Karen Morin in time to include a commentary here on how, in the course of its forensic inquiry into the diverse relationships between animal and carceral spaces (Morin, 2018), she draws startling parallels between the slaughtering of animals and the execution of prisoners (in certain US prisons). Her work, brokering between the concerns of animal and carceral geographers while cross-coding issues of enclosure, slaughter and routine maltreatment of *both* animals and prisoners, offers a complementary account to our own briefer remarks in the present chapter.

References

Adorno, T. W. (1973). *Negative dialectics* (trans.). New York: The Seabury Press.
Agamben, G. (1998). *Homo sacer: Sovereign power and bare life* (trans.). Redwood City, CA: Stanford University Press.
Agamben, G. (1999). *Remnants of Auschwitz: The witness and the archive* (trans.). New York: Zone Books.
Agamben, G. (2005). *State of exception.* Chicago: University of Chicago Press.
American Humane Association. (1908). Report of the Proceedings of the Thirty-second Annual Meeting of The American Humane Association. New Orleans, LA. https://hdl.handle.net/2027/uiug.30112042803137.
Atkins, P. (Ed.). (2012a). *Animal cities: Beastly urban histories.* London: Routledge.
Atkins, P. (2012b). Animal wastes and nuisances in nineteenth-century London. In P. Atkins (Ed.) *Animal cities: Beastly urban histories* (pp. 53–76). London: Routledge.
Atkins, P. (2012c). The urban blood and guts economy. In P. Atkins (Ed.) *Animal cities: Beastly urban histories* (pp. 77–106). London: Routledge.
Ayling, R. S. (1908). *Public abattoirs: Their planning, design, and equipment.* E. & FN Spon.
Blecha, J. L. (2007). *Urban life with livestock: Performing alternative imaginaries through small-scale livestock agriculture in the United States.* Doctoral Diss., retrieval from ProQuest Publication No.3273113, Department of Geography, University of Minnesota.

Blecha, J. L. (2015). Regulating backyard slaughter: Strategies and gaps in municipal live-stock ordinances. *Journal of Agriculture, Food Systems and Community Development* 6, 33–48.

Blecha, J. L. & Davis, A. (2014). Distance, proximity and freedom: Identifying conflicting priorities regarding urban backyard livestock slaughter. *Geoforum* 57, 67–77.

Blecha, J. L. & Leitner, H. (2014). Reimaging the food system, the economy and urban life: New urban chicken-keeping in US cities. *Urban Geography* 35, 86–108.

Blue, G. & Alexander, S. (2017). Coyotes in the city: Gastro-ethical encounters in a more-than-human world. In K. Gillespie & R.-C. Collard (Eds.) *Critical animal geographies: Politics, intersections and hierarchies in a multispecies world* (pp. 149–163). London: Routledge.

Broadway, M. J. (2000). Planning for change in small towns or trying to avoid the slaugh-terhouse blues. *Journal of Rural Studies* 16, 37–46.

Broadway, M. J. & Stull, D. D. (2006). Meat processing and Garden City, KS: Boom and bust. *Journal of Rural Studies* 22, 55–66.

Broadway, M. J. & Ward, J. (1990). Recent changes in the structure and location of the US meatpacking industry. *Geoforum* 75, 76–79.

Brody, D. (1964). *The butcher workmen: A study of unionization*. Cambridge, MA: Harvard University Press.

Brooke, C. (1987). Sacred slaughter: The sacrificing of animals at the Hajj and Id al-Adha. *Journal of Cultural Geography* 7, 67–88.

Buller, H. & Morris, C. (2003). Farm animal welfare: A new repertoire of nature-society relations or modernism re-embedded? *Sociologia Ruralis* 43, 216–237.

Buller, H. & Roe, E. (2012). Commodifying animal welfare. *Animal Welfare* 21, 131–135.

Buller, H. & Roe, E. (2014). Modifying and commodifying farm animal welfare: The econ-omisation of layer chickens. *Journal of Rural Studies* 33, 141–149.

Buller, H. & Roe, E. (2018). *Food and animal welfare: Producing and consuming valuable lives*. London: Bloomsbury Publishing.

Cash, C. (1907). *Our slaughter-house system: A plea for reform*. London: George Bell.

Claflin, K. (2008). La Villette: City of blood (1867–1914). In P. Y. Lee (Ed.) *Meat, moder-nity, and the rise of the slaughterhouse* (pp. 27–45). Hanover, NH: University Press of New England.

Collard, R.-C. & Gillespie, K. (2017). Introduction. In K. Gillespie & R.-C. Collard (Eds.) *Critical animal geographies: Politics, intersections and hierarchies in a multispecies world* (pp. 1–16). London: Routledge.

Collins, D. S. & Huey, R. J. (2015). *Gracey's meat hygiene*. Chichester: John Wiley & Sons.

Commons, J. R. (1904). Labor conditions in meat packing and the recent strike. *The Quar-terly Journal of Economics* 19, 1–32.

Cronon, W. (1992). *Nature's metropolis: Chicago and the Great West*. Chicago: W. W. Norton.

Curran, M. (2001). Policies of exclusion: A Foucauldian analysis of regulation of industrial hog farming in Kentucky. Paper presented at Annual Meeting of the Association of American Geographers, New York.

Drabenstott, M., Henry, M. & Mitchell, K. (1999). Where have all the packing-plants gone? The new meat geography in rural America. *Federal Reserve Bank of Kansas City Eco-nomic Review* 84(3), 65–82. Available (through DigitalCommons@University_of_Nebraska-Lincoln) at http://digitalcommons.unl.edu/bbrup.

Emerson, R. (1860). Fate. In *The Conduct of Life*, www.emersoncentral.com/fate.htm.

Fields, G. (2003). Communications, innovations and territory: The production network of Swift Meat Packing and the creation of a national US market. *Journal of Historical Geography* 29, 599–617.

Fields, G. (2004). *Territories of profit: Capitalist development and the innovative enterprise of G.F. Swift and Dell Computers*. Redwood City, CA: Stanford University Press.

Fink, D. (1998). *Cutting into the meatpacking line: Workers and change in the rural Midwest*. Chapel Hill, NC: University of North Carolina Press.

Fitzgerald, A. (2010). A social history of the slaughterhouse: From inception to contemporary implications. *Human Ecology Review* 17, 58–69.

Ford, H. & Crowther, S. (1922). *My life and work*. London: William Heineman.

Forshaw, A. & Bergström, T. (1990). *Smithfield: Past and present*. London: Heineman.

Foucault, M. (2003). *'Society must be defended': Lectures at the Collège de France, 1975–1976* (trans.). New York: Picador.

Foucault, M. (2007). *Security, territory, population: Lectures at the Collège de France, 1977–1978* (trans.). London: Palgrave Macmillan.

Foucault, M. (2008). *The birth of biopolitics: Lectures at the Collège de France, 1978–1979* (trans.). London: Palgrave Macmillan.

Furuseth, O. J. (1997). Restructuring of hog farming in North Carolina: Explosion and implosion. *Professional Geographer* 49, 391–403.

Giedion, S. (1948). *Mechanization takes command: A contribution to anonymous history*. New York: Oxford University Press.

Gillespie, K. & Collard, R.-C. (Eds.). (2017). *Critical animal geographies: Politics, intersections and hierarchies in a multispecies world*. London: Routledge.

Giraud, E. (2017). Practice as theory: Learning from food activism and performative protest. In K. Gillespie & R.-C. Collard (Eds.) *Critical animal geographies: Politics, intersections and hierarchies in a multispecies world* (pp. 36–53). London: Routledge.

Heiss, H. (1907). The German abattoir. In C. Cash (Ed.) *Our slaughter-house system* (pp. 85–212). London: George Bell.

Hobson, K. (2007). Political animals? On animals as subjects in an enlarged human geography. *Political Geography* 26, 250–267.

James, B. W. (1880). How abattoirs improve the sanitary condition of cities. *Public Health Papers and Reports* 6, 231–238.

Johnston, C. (2013). Geography, science and subjectivity: Farm animal welfare in the United States and Europe. *Geography Compass* 7, 139–148.

Joyce, J., Nevins, J. & Schneiderman, J. S. (2017). Commodification, violence and the making of workers and ducks at Hudson Valley Foie Gras. In K. Gillespie & R.-C. Collard (Eds.) *Critical animal geographies: Politics, intersections and hierarchies in a multispecies world* (pp. 93–107). London: Routledge.

Lakshmi Singh, A., Jamal, S., Baba, S. A. & Islam, M. M. (2014). Environmental and health impacts from slaughterhouses located on the city outskirts: A case study. *Journal of Environmental Protection* 5, 566–575.

Laxton, P. (2012). This nefarious traffic: Livestock and public health in mid-Victorian Edinburgh. In P. Atkins (Ed.) *Animal cities: Beastly urban histories* (pp. 107–171). London: Routledge.

Lee, P. Y. (2008). Siting the slaughterhouse: From shed to factory. In P. Y. Lee (Ed.) *Meat, modernity, and the rise of the slaughterhouse* (pp. 46–70). Hanover, NH: University Press of New England.

Lorimer, J. (2015). *Wildlife in the Anthropocene: Conservation after nature*. Minneapolis: University of Minnesota Press.

MacLachlan, I. (1998a). George Henry Hammond. In N. L. Shumsky (Ed.) *Encyclopedia of Urban America: The cities and suburbs* (pp. 335–336). New York: ABC-CLIO Publishing.

MacLachlan, I. (1998b). Gustavus Franklin Swift. In N. L. Shumsky (Ed.) *Encyclopedia of Urban America: The cities and suburbs* (p. 769). New York: ABC-Clio Publishing.

MacLachlan, I. (2001). *Kill and chill: Restructuring Canada's beef commodity chain*. Toronto: University of Toronto Press.

MacLachlan, I. (2004). Betting the farm: Food safety, risk society, and the Canadian cattle and beef commodity chain. In A. Heintzman & E. Solomon (Eds.) *Feeding the future: From fat to famine, how to solve the world's food crises* (pp. 40–69). Toronto: House of Anansi Press.

MacLachlan, I. (2005). 'The greatest and most offensive nuisance that ever disgraced the capital of a kingdom': The slaughterhouses and shambles of modern Edinburgh. *Review of Scottish Culture* 17, 57–71.

MacLachlan, I. (2007). A bloody offal nuisance: The persistence of private slaughterhouses in nineteenth-century London. *Urban History* 34, 226–253.

MacLachlan, I. (2008). Humanitarian reform, slaughter technology, and butcher resistance in nineteenth-century Britain. In P. Y. Lee (Ed.) *Meat, modernity, and the rise of the slaughterhouse* (pp. 107–126). Hanover, NH: University Press of New England.

Miele, M. & Evans, A. (2010). When foods become animals: Ruminations on ethics and responsibility in care-*full* practices of consumption. *Ethics, Place and Environment* 13, 171–190.

Miller, K. (1956). The economic impact of technology on meat packing. *Journal of Farm Economics* 38, 1775–1778.

Morin, K. M. (2018). *Carceral space, prisoners and animals*. London: Routledge.

Mundigo, A. I. & Crouch, D. P. (1977). The city planning ordinances of the laws of the Indies revisited: Part I: Their philosophy and implications. *Town Planning Review* 48, 247–268.

Otter, C. (2006). The vital city: Public analysis, dairies and slaughterhouses in nineteenth-century Britain. *Cultural Geographies* 13, 517–537.

Otter, C. (2008a). Civilizing slaughter: The development of the British public abattoir, 1850–1910. In P. Y. Lee (Ed.) *Meat, modernity, and the rise of the slaughterhouse* (pp. 89–106). Hanover, NH: University Press of New England.

Otter, C. (2008b). *The Victorian eye: A political history of light and vision in Britain, 1800–1910*. Chicago: University of Chicago Press.

Page, B. (1998). Rival unionism and the geography of the meatpacking industry. In A. Herod (Ed.) *Organizing the landscape: Geographical perspectives on labor unionism*. Minneapolis: University of Minnesota Press.

Perren, R. (1978). *The meat trade in Britain, 1840–1914*. London: Routledge and Kegan Paul.

Philo, C. (1995). Animals, geography and city: Notes on inclusions and exclusions. *Environment and Planning D: Society and Space* 13, 655–681.

Philo, C. (1998). Animals, geography and city: Notes on inclusions and exclusions. In J. Wolch & J. Emel (Eds.) *Animal geographies: Place, politics and identity in the nature-culture borderlands* (pp. 51–70). New York: Verso [an abridged version of Philo 1995].

Philo, C. (2017). Squeezing, bleaching and the victims' fate: Wounds, geography, poetry, micrology. *GeoHumanities*, on-line preview at http://dx.doi.org/10/1080/23735 66X.2017.1291311.

Racine Jacques, J. (2015). The slaughterhouse, social disorganisation and violent crime in rural communities. *Society and Animals* 23, 594–612.

Rasmussen, C. (2017). Pleasure, pain and place: Ag-gag, crush videoa and animal bodies on display. In K. Gillespie & R.-C. Collard (Eds.) *Critical animal geographies: Politics, intersections and hierarchies in a multispecies world* (pp. 54–70). London: Routledge.

Roe, E. (2010). Ethics and the non-human: The matterings of animal sentience in the meat industry. In B. Anderson & P. Harrison (Eds.) *Taking place: Non-representational theories and geography* (pp. 261–282). London: Ashgate.

Roe, E. (2013). Global carcass balancing: Horsemeat and the agro-food network. *Radical Philosophy* 179(May/June), 2–5.

Roe, E. (2016). Meat-packing. In J. Urbanik & C. Johnston (Eds.) *Humans-animals: A geography of co-existence* (pp. 239–240). Santa Barbara, CA: ABC-CLIO Publishing.

Ruttan, V. W. (1954). *Technological progress in the meatpacking industry, 1919–1947.* Washington, DC: United States Department of Agriculture.

Schwarz, O. (1901). *Public abattoirs and cattle markets*. Ice and Cold Storage Publishing Company.

Smith, D. C. & Bridges, A. E. (1982). The Brighton market: Feeding nineteenth-century Boston. *Agricultural History* 56, 3–21.

Stull, D. D. & Broadway, M. J. (2012). *Slaughterhouse blues: The meat and poultry industry in North America*. Belmont: Wadsworth Cengage Learning.

Stull, D. D., Broadway, M. J. & Griffith, D. (1995). *Any way you cut it: Meat processing and small-town America*. Lawrence, KS: University Press of Kansas.

Trabsky, M. (2015). Institutionalising the public abattoir in nineteenth-century colonial society. *Australian Feminist Law Journal* 40, 169–184.

Ufkes, F. M. (1995). Lean and mean: US meat-packing in an era of agro-industrial restructuring. *Environment and Planning D: Society and Space* 13, 683–706.

Ufkes, F. M. (1998). Building a better pig: Fat profits in lean meat. In J. Wolch & J. Emel (Eds.) *Animal geographies: Place, politics and identity in the nature-culture borderlands* (pp. 241–255). New York: Verso.

Urbanik, J. (2012). *Placing animals: An introduction to the geography of human-animal relations*. Lanham: Rowman & Littlefield.

Walsh, M. (1978). The spatial evolution of the Midwestern pork industry, 1835–1875. *Journal of Historical Geography* 4, 1–22.

Warren, W. J. (2007). *Tied to the great packing machine: The Midwest and meatpacking*. Iowa City, IA: University of Iowa Press.

Watts, M. (2000). Afterword: Enclosure. In C. Philo & C. Wilbert (Eds.) *Animal spaces, beastly places: New geographies of human-animal relations* (pp. 292–304). London: Routledge.

Watts, M. (2004). Are hogs like chickens? Enclosure and mechanizatio in two 'white meat' filières. In A. Hughes & S. Reimer (Eds.) *Geographies of commodity chains* (pp. 39–62). London: Routledge.

White, R. J. (2017). Animal geographies, anarchist praxis and critical animal studies. In K. Gillespie & R.-C. Collard (Eds.) *Critical animal geographies: Politics, intersections and hierarchies in a multispecies world* (pp. 19–35). London: Routledge.

Wolch, J. & Emel, J. (Eds.). (1995). Bringing the animals back in: Special issue. *Environment and Planning D: Society and Space* 13.

Wolch, J. & Emel, J. (Eds.). (1998a). *Animal geographies: Place, politics and identity in the nature-culture borderlands*. New York: Verso.

Wolch, J. & Emel, J. (1998b). Preface. In J. Wolch & J. Emel (Eds.) *Animal geographies: Place, politics and identity in the nature-culture borderlands* (pp. xi–xx). New York: Verso.

8   The pigs are back again

Urban pig keeping in wartime Britain, 1939–45

Thomas Webb

In April 1943, the popular British pictorial magazine *Picture Post* reported on the voluntary efforts of members of the National Fire Service, who formed a pig club within the ruins of bombed London. Titled "Pigs in the Ruins" (1943, p. 18), the report described that:

> The pigs – who rooted the soil of London when London soil was sweet – are back again on the old farmlands just outside of the city walls. The buildings – which rose out of the pastures – are lying smashed and deserted. The ground – usurped by lawyers, surgeons, estate agents, and what not – has reverted (temporarily, at any rate) to the original tenants of the land.
>
> On a casual walk round the Lincoln's Inn Fields you wouldn't notice any difference at all. A row of fire appliances half conceals the entrance to the piggery. The ruins of the College of Estate Management – and the Museum wing of its next-door neighbour, the Royal College of Surgeons – show no outward sign that new occupants have moved in.
>
> But if you peer through the gaping brickwork – into the cavernous spaces where, once, the surgeons used to pickle bits of you and me as exhibits and the estate agents contrived to baffle us with the intricacies of property law – your eyes will be gratified by the pink backs of ranging porker. They are the charges of the men of the National Fire Service whose appliances stand on the kerb.

As this extract indicates, pigs were reintroduced into areas of wartime London that had long been absent of livestock animals, including bombed ruins of former city buildings. They were reared in areas that were close to sources of kitchen waste for pigswill, which was a vital resource amid wartime food import restrictions. Pigs were subsequently utilized, as part of wider recycling efforts in wartime Britain, to convert kitchen waste into bacon.[1] But whilst pigs were reintroduced into areas of wartime London, as well as other urban areas in Britain, they were also hidden within these landscapes. The journalist noted how it was easy to be oblivious to the piggery unless you peered through the brickwork, where the pigs were enclosed in the former yard of the Royal College of Surgeons. This chapter charts the processes by which pigs reoccupied spaces within wartime British towns and cities. This process was shaped by historical anxieties of pigs within urban areas, ensuring that they were both included and excluded from such spaces during the war. It was also contingent on the pigs' ability to survive within such environments, and one in which the ultimate slaughter of the pigs redefined where animal death could appropriately take place. This chapter explores how the demands of World War II temporarily redefined the place, and influence, of the pig within British towns and cities.

The utilization of pigs within British towns and cities was contingent on the formation of pig clubs, such as that of the firemen's pig club in the bombed ruins of London. These clubs were underpinned by the formation of the Small Pig Keeper's Council (SPKC) in 1940, which was created by the government to oversee the development of pig clubs on a national scale. With campaigns such as a "Pig on Every Street," pigs were kept on waste grounds, gardens, allotments, and smallholdings (SPKC Annual Report, 1942). In urban and suburban areas, dustmen, policemen, firemen, transport workers, civil defense station workers, and canteen staff, amongst others, established new co-operative clubs under the guidance of the SPKC. These clubs kept pigs on a shared site, and members contributed through maintaining the pigsties and feeding the pigs. They were primarily fed on salvaged kitchen scraps, which had been boiled down and turned into pigswill. By March 1945, the movement had expanded significantly, with 6,900 SPKC affiliated clubs owning approximately 140,000 pigs, spread across the length and breadth of the British Isles, incorporating individuals from varying regions and backgrounds (SPKC Annual Report, 1945).

The success of these urban pig clubs involved the formation of new animal spaces. This concept, proposed by Chris Philo and Chris Wilbert (2000), offers a fruitful way to explore how the totalizing tendencies of World War II re-spatialized animals in both physical and imaginative terms. A focus on animal spaces involves explorations of the discursive construction of animals, including how they were categorized, the spaces that were allocated to them, and where they were seen as in or out of place. Recognizing that animals have been socially defined, it requires an examination of the various ways in which animals were placed by human societies in their local material spaces (such as domestic homes, fields, or factories) and in a range of imaginary or cultural spaces (Philo & Wilbert, 2000). Paying attention to the construction of animal spaces is productive in this context, as pigs were physically moved into spaces that prior to the war would have been out of place. Likewise, it also involved the imaginative placement of animals as either allies or enemies of the war effort, which had material effects on the ways in which pigs were harnessed in regards to wartime recycling and food production demands. In light of this, focusing on animal spaces reveals the impact and legacy of the war on the physical and imaginative place of pigs within Britain.

The successful operation of pig clubs was reliant on being close to sources of kitchen waste. This meant that pig clubs were frequently formed in urban and suburban spaces that were near canteen, restaurant, and domestic home kitchens. Subsequently, the first section of this chapter examines how pigs were physically repositioned in urban spaces. The focus on the rearing of livestock in urban settings redresses the historiographical balance of British wartime farming, which is predominantly focused on the countryside (Short, Watkins & Martin, 2007). Pigs, chickens, and rabbits were reared on a much greater scale than before within city allotments, yards, gardens, and waste spaces, as part of the "Grow More Food" campaign (Strang, 1939). This expansion was significant because it countered the sanitization of the British cityscape as a livestock-free space that occurred over the course of the second half of the nineteenth century (Atkins, 2012). Technical, sanitized, and visual changes wrought by the introduction of the abattoir normalized the private act of slaughter and contributed toward the exclusion of livestock from most areas of the city (Otter, 2006; Otter, 2008). Moreover, the association of urban livestock with the slum dweller, and the "dangerous" spaces they cohabited, ensured the removal of livestock formed an integral part of the "civilizing" mission for nineteenth-century city reformers (Atkins, 2012). In light of such associations, pigs were perceived to be dangerous animals. As Robert Malcolmson and Stephanos Mastoris (1998, pp. 1–2) note in regards to the inter-war period, pigs had become a "common-place metaphor for human waste and disgust, and . . . a metaphorical association with untidiness, disorder and filth," and served to "define in consciousness a boundary between civilised and the uncivilised, the refined and the unrefined." The majority of the urban populace could not challenge this image through engagement with actual pigs as inter-war production was primarily centered on indoor factory-style sites or on rural farms and smallholdings – the pig only entered the city via the abattoir and butchers' shops (Woods, 2012).

The extensive reintroduction of pigs into the urban sphere required the symbolic and material boundaries – which defined humans and animals as distinct categories, and which separated the sanitary city from rural nature – to be redrawn (Philo, 1995). The occurrence of this process in British cities during World War II demonstrates the power of war for shaping the uses of the urban environment, and for redefining the boundaries of inclusionary and exclusionary practices, with reference to animals. Furthermore, the transitory nature of this process of reintroduction within the war years indicates not only the significance, but also the distinctiveness of this short period, amidst the wider history of urban human–pig relations (McNeur, 2011).

Nevertheless, the physical movement of pigs into urban spaces was not always a smooth process. The utilization of pigs during the war was contingent on how they both allowed and blocked, or enabled and thwarted, these processes. As Chris Pearson (2015) highlights, it is problematic to consider that animals have "collaborated" with, or "resisted" against, humans in the past. Those who utilize such language in regards to animals often do so in an attempt to grant them agency within the frameworks of social history and histories of capitalism to uncover the "resistance" or "labour" of the animal (Hribal, 2007). Instead, focusing on how animals have allowed and/or blocked, or enabled and/or thwarted, historical processes allows an examination of animal agency beyond anthropocentric notions of subjectivity, intention, or reason (Pearson, 2015). For the case of pigs in urban pigsties during World War II, the cramped conditions of some urban spaces led to outbreaks of swine-related diseases, which hampered how they were bred and reared. As such, the second section of this chapter examines outbreaks of various diseases in pigs within wartime British cities. This reveals how pigs and other nonhumans blocked (or thwarted) pig keepers' attempts to rear their animals within urban environments. In response, urban pig keepers adapted to the "natural" needs of the pigs.

The final section of this chapter looks at the end of the pig's utilization into the war effort through an analysis of their slaughter and conversion into consumable meat. This intersects with longer-term historical debates surrounding slaughter, cultural sensibilities, and the removal of tangible signs of death within urban spaces, by uncovering how slaughtering practices were extended into visible locations for the needs of war (Otter, 2006; Otter, 2008; Fitzgerald, 2010). This demonstrates how the process of including pigs within cities not only enabled urban citizens to interact with these animals, but also exposed them to their deaths. It also highlights how the circumstances of war redefined the spaces of legitimate and illegitimate livestock slaughter.

Inclusions and exclusions within the city

As Rauno Lahtinen and Timo Vuorisalo (2004) show in relation to the Finnish city of Turku during World War II, the needs of war fundamentally changed how people experienced and used the urban environment. Agricultural animals were reintroduced for food production purposes, and the spaces that animals were allowed to occupy were redefined accordingly. Urban wastelands, gardens, parks, yards, and allotments were appropriated for pig keeping in wartime Britain. This process of reintroduction raises important questions regarding the constitutive process by which animals are included and excluded from certain spaces, and how these boundaries are shaped and contested within particular cultural contexts.

Chris Philo (1995) argues that we need to recover how human communities think, feel, and talk about the animal in question. How the animal is understood and conceptualized will shape the "sociospatial practices towards these beings on an everyday basis," and this has important consequences for whether the particular species of animal is included or excluded from common sites of human activity (Philo, 1995, p. 656). In the context of the wartime urban environment in Britain, the pig was redefined as an efficient food producer and waste converter in official rhetoric. Combined with the exigencies of war, and the availability of kitchen refuse within towns and cities, the pig was allowed to re-enter the urban sphere on a far greater scale than it had prior to the war. This helped change the urban environment from an exclusionary to an inclusionary space for the pig. Nevertheless, historical anxieties about the pig within urban spaces remained.

Legislation relating to urban pig keeping was altered to ensure that the urban environment was legally made an inclusionary space for pigs. Pre-war urban pig keeping had been hindered by tenancy restrictions and local bylaws, which were extended under the Public Health Act 1936. The locality of these restrictions meant they could vary in scale and specificity within different locales, although the Ministry of Health did offer guidelines that pigs needed to be kept at least 100 feet from residential dwellings in cleanly maintained sties that were located in areas which would not threaten water pollution. These restrictive measures were partially overturned in an Order of Council passed on 19 June 1940. A response to a query from a potential pig keeper in Preston by a Ministry of Health spokesman summarized this order succinctly:

> it is lawful for any person to keep pigs, hens or rabbits in any place notwithstanding any lease or tenancy or other contract, or any restriction imposed by or under any enactment; so long as the keeping of such pigs, hens or rabbits is not in fact prejudicial to health or a nuisance.

> (Shelley, 1940, p. 1)

Whilst the urban environment was opened up legally to the pig, historical anxieties associating urban agriculture with public health and nuisance still pervaded over the official redefinition of the pig as useful for the war effort. Even toward the end of the war, proponents of pig keeping were still tackling the "persistent problem" of urban authorities that regarded the pig as "not a clean animal" who "makes his presence felt some distance round his actual quarters" (Ministry of Agriculture Memoranda, 1944, p. 1).

Pig clubs negotiated this conflicting definition of the pig as both useful and a threat – in terms of health and nuisance – by hiding pigs within the urban landscape. In this way, the pig was both included and excluded from sites of common human activity within towns and cities. For instance, the pigs at Nine Elms Police Station were made unnoticeable, as buildings and a railway track surrounded the pigsty. Prior to the club's formation, a Medical Officer of Health inspected the site and granted permission on the condition that it would not disturb, or be visible to, local residents, despite its close proximity to residential dwellings only thirty feet away (Bartell, 1941). Firemen, factory workers, A.R.P. depot workers, and dustmen, who set up pig clubs in the borough of Tottenham, took similar precautions and made certain that their pigsties were located on sites away from main buildings and other premises (Blair, 1943). Scholars argue that these kinds of boundaries are ordered to maintain the distinction between urban civilization and animal nature. They stop the threat of the animal transgressing into the sanitary city, which helps maintain the distinction between human and animal, and order and disorder (Griffiths, Poulter & Sibley, 2000). Evidently, the pig threatened this boundary in wartime Britain and was therefore hidden from sight.

Nevertheless, some pig clubs openly exposed this boundary, and they were subsequently monitored and regulated through sanitization. For instance, the policemen members of a club located on the Frame Ground in Hyde Park were expected to maintain the area's cleanliness, prevent any pollutant smells, keep noise to a minimum, and erect fences to ensure pigs did not escape (Department of Works Memoranda, 1941). Moreover, it was only the perceived propaganda benefits of such a high-profile voluntary club that ensured its approval, and government officials stated that such an endeavor would not have been entertained in peacetime. This highlights that club members were expected to maintain and display a sanitized space, and that the visible pig was considered a pollutant that threatened the urban environment. As Mary Douglas argues, pollutants (dirt) are considered as "matter out of place" and are culturally specific in relation to time and space (Douglas, 2002; Smith, 2007). The labelling of dirty and clean can function to establish social order, and cleansing practices can be employed to enforce regulatory and exclusionary practices within communities (Cooper, 2010). The process of sanitization then served as an apparatus for monitoring and regulating the perceived threat of the pig transgressing into exclusionary spaces, and reinforced the material and symbolic boundaries between the sanitized city and the perceived threat of animal nature.

Outbreaks of disease

Sanitizing urban pig spaces also served the purpose of tackling livestock and swine-related diseases. During the early years of the war, cases of foot-and-mouth disease, swine erysipelas, swine fever, and necrotic enteritis broke out amongst pig populations across both urban and rural areas of the country ("In Parliament – Foot-and-Mouth Disease," 1940; "Diagnosis of Diseases of Swine," 1940). Whilst it cannot be determined that these problems were caused by the reintroduction of pigs into towns and cities alone, discussions by government officials, the Public Cleansing Institute, and veterinarians suggest the conditions in which pigs were kept in these spaces were perceived to be highly contributory (A.L.T., 1941). Furthermore, the widespread use of kitchen-refuse-formed-pigswill was perceived as integral to these disease-related issues. Combined with the cramped locations of urban pigsties, not only was disease an issue, but the physiological health of the pig was also inhibited (Blair, 1943). Therefore, diseases and the physical impairment of pigs constrained human activity in the form of food production. But the human responses to these problems also indicate that nonhuman factors shaped human intentions, both in terms of how they viewed and treated the pig, and through how they interacted with the urban environment to amend these problems.

As the war progressed, commentators began to discuss the detrimental effects of wartime pigswill on the health of the pig, as pigswill acted as a vector for spreading disease. Veterinarians observed how a change from a diet largely consisting of concentrates to one that was based from swill produced primarily from kitchen refuse led to worms and digestive conditions (Woods, 2012). Pigswill was also connected with the spread of foot-and-mouth disease. During 1942, forty-one initial outbreaks led to a further 629 over the course of the year, and the disease's potential to devastatingly impact agricultural animals through cross-species transmission ensured it received critical wartime attention (Moore, 1945). Members of Parliament discussed the processes by which pigswill became contaminated, and laid blame on imported meat from South America – the bones of which were added to pigswill ("In Parliament – Foot-and-Mouth Disease," 1940). This infected pigswill could then contaminate other swill stored nearby. Butchers who came into contact with contaminated meat also spread the disease, via infected blood on their smocks when they visited local pigsties. Moreover, parliamentarians and veterinarians identified how dogs helped spread the disease on several occasions by carrying contaminated bones into spaces occupied by pigs ("In Parliament – Foot-and-Mouth Disease," 1944).

The cramped conditions of many urban pigsties were also detrimental to the health of the pigs. Veterinarians predicted in early 1940 that the pre-war indoor intensively farmed pig, which had a high rate of weight gain, had acquired this capacity at the expense of disease resistance. The shift to urban allotments or wastelands would expose pigs to disease and, subsequently, they would require a number of years to build up a resistance to these environments ("Diagnosis of Diseases of Swine," 1940). This prediction appears to have proved correct, especially in veterinarian J. S. Blair's (1943) case study of "The Problems of Pig Clubs in War-time," which focused upon the district of Tottenham. Blair outlined various problems that urban pig keepers faced within his North London district. These included the problems of rearing piglets between the age of weaning and three-and-a-half months on premises, where the pig enclosure had no extra grass or earth available for movement, which caused these piglets to be inflicted with a "thick unhealthy skin, a tucked-up, unthrifty, anaemic appearance, and rickets" (Blair, 1943, p. 445). Problems also arose concerning the heating and ventilation of urban pig enclosures, and the use of salvaged sawdust as bedding, which got into troughs and was consumed, causing digestion problems. These congestive enclosures led to outbreaks of swine erysipelas amongst older pigs, necrotic enteritis amongst piglets, and one outbreak of swine fever in the district – although Blair accredits the limited impact of swine fever to the lack of trade outside of the district and efforts to sanitize and maintain pig enclosures.

Throughout the war, efforts were made to combat these threats to the pigs' health whenever possible. For instance, the Foot-and-Mouth Disease Act was amended in 1941 with a "Boiling of Animal Foodstuffs" Order, which stated that all swill must be boiled for over an hour before it was served as food (Worden, 1943). This would, in theory, alleviate problems of disease and digestion. Public cleansing officials also urged dustmen to familiarize themselves with foot-and-mouth disease orders and become "sanitary conscious" in response to the threat of contaminated materials (A.L.T., 1941). In relation to the problems of pig keeping in Tottenham, Blair (1943) demonstrated that by 1943 many of the issues faced were substantially rectified. Pig producers alleviated the problems caused by sawdust through the use of non-harmful straw for bedding. The issues associated with ventilation within urban pig enclosures were substantially improved by the trial-and-error of various designs, which sought greater access to fresh air and soil for improvement. In particular, "the outdoor sty with outdoor run on earth" was seen as the most beneficial type of sty for farrowing and for alleviating the detrimental impact of the urban environment on piglets (Blair, 1943, pp. 45–46). This sty allowed these piglets to root and perform natural outdoor tendencies, which aided health through exercise. Furthermore, if the location did not permit the rearing of piglets, clubs bought pigs once they had grown to a size where the lack of soil (or nature) would not be as detrimental.

Such responses to urban pig keeping corroborate Abigail Woods' (2012) argument that during the first half of the twentieth century, some pig producers believed that "nature" was essential for successful pig production. These commentators voiced concerns with pre-war "artificial" methods of agriculture and moralized about the danger of breaking nature's laws and, therefore, tried to make the pig's environment as close to rural nature as possible. The human responses to the non-human problems of urban pig keeping show how the pig's environment – through improved ventilation, an extension of enclosures, and by other means – was naturalized, and the biological needs of the pig were recognized in this process. This extends Woods' analysis, as her argument does not take into account the circumstances of the urban environment. Moreover, the examples of the urban environment nuance her argument as they show how the naturalization of these pig enclosures was coupled with a process of "artificial" sanitization. In other words, urban pig enclosures became hybrid spaces whereby the city met the countryside. They were also spaces in which humans responded to, and worked with, nonhuman factors that potentially hampered efficient pig production. Through working with the needs of the pig, the pig club movement was more effective, and the nation's food stocks were enhanced.

Slaughter, sensibilities, and supply

Once successfully reared, the final stage of the pigs' enrollment was their slaughter and subsequent conversion into bacon and pork. Like the tentative reintroduction of pigs into towns and cities, the laws relating to animal slaughter were redefined to suit the needs of war to ensure that, in theory, pig owners contributed to the nation's food supply and were kept within the allowances of rationing. Critical to this process was acquiring a slaughter license from either a Food Executive Officer or an Area Meat and Livestock Officer (Calder, 1945). For individual self-suppliers, licenses for slaughter were granted to applicants for no more than two pigs per household per year. The applicant must have also reared the pig for at least three months and was required to surrender fifty-two bacon coupons for each pig or sell one whole side of the pig to the Ministry of Food. For pig clubs, slaughter licenses were granted on the basis that the number of pigs slaughtered did not exceed the equivalent of two per member per year. The meat was only allowed to be distributed amongst members, and for every pig slaughtered another had to be sold to the government via a Collecting Centre or Pig Allocation Officer. To aid these processes, licenses also permitted slaughter to take place on an owner's premises, or any other premises approved by a Food Executive Officer, so long as it was permitted by local regulations and bylaws.

This redefinition of where slaughter could take place offers a glimpse into how the war disrupted longer-term changes regarding human attitudes and policies toward animal slaughter and food production within towns and cities. As historiography suggests, over the second half of the nineteenth century, sanitary discourses combined with expanding animal geographies, globalized networks of trade, and the industrialization of slaughter, to ensure that agricultural animals were "visibly" removed from the urban sphere (Otter, 2006; Philo, 1995). This timeframe is a fault line in the long history of meat production, where industrialization distanced people from what they consumed, the act of killing, and the natural environment in which animals were raised (Fitzgerald, 2010). Chris Otter (2008, p. 103) argues that "the beginning of the twentieth century saw the completion of the psychological mechanism that removed death from society, eliminated its character of public ceremony, and made it a private act" in regards to animal slaughter in Britain. Instead, in early twentieth-century urban and suburban Britain, meat had been commoditized and sanitized through its point of access (the butcher's shop), through inspection, and through the way it was displayed (Otter, 2008). This process, in turn, set conditions by which true disgust could be felt when faced with the act of animal slaughter itself or at least certain sensorial engagements, such as the sight of blood (Otter, 2008).

Although it is difficult to trace how citizens reacted to the slaughter of pigs in urban and suburban yards and smallholdings, SPKC instruction manuals guided pig club members on how to slaughter and cure their pigs in their own yards (Ministry of Agriculture and Small Pig Keeper's Council, 1943). These were framed as methods that, due to interwar regulations, were in danger of dying out and were in need of a revival (Ministry of Agriculture and Small Pig Keeper's Council, 1943). Similarly, the wartime British national press reported on schemes whereby girls and young women were trained in Lancaster to cure bacon at home (Training to Save the Bacon, 1943). These official practices of slaughter at home were also accompanied by cases of illegal slaughter – where the pig owner had not been granted a slaughter license – for the distribution of meat within local communities and on the black market (Roodhouse, 2013). The slaughter of pigs, and the subsequent curing of bacon, were then reframed during the war as acts that were normalized – in both legal and illegal instances – within the urban backyard, smallholding, and home.

Slaughter and meat preparation did not always occur on owners' premises, however, and the systems that corporeally and psychologically removed death from society were still maintained in parallel. Regulations ensured that pigs were also sold or donated to the Ministry of Food. In these instances, pork pigs were sent to one of the 611 government slaughterhouses in operation, and bacon pigs were sent to one of the 140 bacon factories operated by private enterprise but regulated by ministry officials. From the hidden or marginal locations of these institutions pigs were slaughtered, converted into bacon or pork, and passed through wholesalers, to retailers, and finally to rationed consumers. This system was in place for the majority of the war, although bacon was not rationed until January 1940, and all other meat followed in March of the same year (Hammond, 1962). It is within this system that a large quantity of wartime bacon and pork circulated within towns and cities.

Conclusion

The end of World War II in 1945 did not see the end of British urban pig keeping. Within the immediate post-war constraints of a global food shortage, British citizens were still subjected to government policies of rationing and austerity. The continued rationing of meat until 1954 ensured that the wartime pig club movement grew throughout the later 1940s, before declining significantly in the early mid-1950s. This continued post-war growth was enabled by the extension of wartime legislation post-1945, which allowed pig clubs to continue operating in spaces where pig keeping may have otherwise been restricted. Regulation 62B of the Defence (General) Regulations, 1939, which suspended restrictions on the keeping of pigs, hens, and rabbits by tenants and occupiers of land, was extended until July 1951. Section 12 of the Allotments Act, 1950, subsequently replaced this legislation, and continued the suspension of restrictions on the keeping of hens and rabbits, but no longer covered pigs. Instead, as a circular from Whitehall to local authorities noted, it was now within the discretion of local authorities to decide whether it was suitable to keep pigs on their housing estates. But as the circular stated in support of continued pig keeping:

> In dealing with applications of this kind local authorities will no doubt bear in mind that it is still necessary in the national interest to encourage domestic food production and will not withhold their consent if it can be given without detriment to the maintenance of satisfactory standards of amenity and public health.
>
> (Hutchinson, 1951, p. 2)

This support for pig keeping by the British state, however, was not always reciprocated by local authorities post-1945. Mirroring the tensions between domestic pig keepers during the war, pre-war anxieties associating urban pig keeping with public health and nuisance still pervaded. Subsequently, the SPKC were continually forced to defend their members, who were summoned to court under the Nuisances provisions of the Public Health Act, 1936 (SPKC Annual Report, 1946), against the decisions of local authorities. The regular success of the SPKC in defending their members in court demonstrates that urban pig keeping was perceived to be in the national interest, even if it remained a site of tension between pig club members and local authorities surrounding historical anxieties concerning pigs and questions of public health.

The presence of pigs within British towns and cities from the early 1940s until the early 1950s highlights how the demands of World War II temporarily redefined the urban sphere as a space for pigs and a legitimate site for domestic livestock slaughter. While the place of pigs and domestic slaughter was a source of tension and debate, especially between pig keepers and local authorities, their presence nonetheless complicates arguments that livestock was visibly removed from British towns and cities from the late nineteenth century onwards (Atkins, 2012). Rather, it indicates the need for further temporally and spatially specific case studies for uncovering the place of pigs – and livestock more generally – within twentieth-century histories and geographies of British and other western urban spaces. Moreover, the ability of pigs – through outbreaks of disease – to block, thwart, and/or hamper their return to the urban environment stresses the need to pay further attention to how animals survive within cities, and how humans interact and respond to their needs.

Note

1 For histories of war and recycling in Britain, see: Cooper (2008); Riley (2008); Thorsheim (2013); Thorsheim (2015); Irving (2016).

References

A.L.T. (1941). Our Concern with the Diseases of Animals Act. *Public Cleansing and Salvage, 31*(372), 380–381.

Atkins, P. (2012). Animal Wastes and Nuisances in Nineteenth Century London. In P. Atkins (Ed.). *Animal Cities: Beastly Urban Histories* (pp. 19–52). Farnham: Ashgate.

Bartell, G. (1941, June 16). *Letter from G. Bartell (S.D. Inspector of Nine Elms Police Station) to Superintendent.* Police Pig and Poultry Clubs – Rules and Regulations (1941; 1952) (MEPO 2/6516). National Archives, London, United Kingdom.

Blair, J. S. (1943). The Problems of Pig Clubs in War Time. *Veterinary Record, 46*(55), 445–448.

Calder, A. (1945). *Ministry of Food Regulations for Issue of Licences to Slaughter to Self-Suppliers of Pigs (1945).* Wartime Pig Control – An Outline by Dr A Calder (1945) (MAF 223/53). National Archives, London, United Kingdom.

Cooper, T. (2008). Challenging the 'Refuse Revolution': War, Waste and the Rediscovery of Recycling, 1900–1950. *Historical Research, 81*(214), 710–731.

Cooper, T. (2010). Recycling Modernity: Waste and Environmental History. *Historical Compass, 8*(9), 1114–1125.

Department of Works Memoranda. (1941). *Memoranda about Pigs in Frame Ground in Hyde Park, 11/7/1941.* Police Station Pig Club in the Frame Ground (1940–1955) (WORK 16/1562). National Archives, London, United Kingdom.

Diagnosis of Diseases of Swine. (1940). *Veterinary Record, 12*(52), 224.

Douglas, M. (2002). *Purity and Danger: An Analysis of Concepts of Pollution and Taboo* (6th ed.). London: Routledge.

Fitzgerald, A. (2010). A Social History of the Slaughterhouse: From Inception to Contemporary Implications. *Human Ecology Review, 17*(1), 58–69.

Griffiths, H., Poulter, I., & Sibley, D. (2000). Feral Cats in the City. In C. Philo & C. Wilbert (Eds.). *Animal Spaces, Beastly Places: New Geographies of Human-Animal Relations* (pp. 56–70). London: Routledge.

Hammond, R. J. (1962). *Food, Volume III: Studies in Administration and Control.* London: Her Majesty's Stationery Office and Longmans, Green and Co.

Hribal, J. C. (2007). Animals, Agency, and Class: Writing the History of Animals from Below. *Human Ecology Forum, 14*(1), 101–112.

Hutchinson, N. (1951, August 28). *Circular Notice from N. Hutchinson to all Housing Associations, Development Corporations and County Councils in England.* Keeping of Pigs, Poultry and Rabbits on Local Authorities' Housing Estates: Papers Leading Up to Draft Circular (1951–1955) (HLG 101/310). National Archives, London, United Kingdom.

In Parliament – Foot-and-Mouth Disease. (1940). *Veterinary Record, 12*(52), 30.

In Parliament – Foot-and-Mouth Disease. (1944). *Veterinary Record, 48*(56), 463.

Irving, H. (2016). Paper Salvage in Britain during the Second World War. *Historical Research, 89*(244), 373–393.

Lahtinen, R., & Vuorisalo, T. (2004). 'It's War and Everyone Can Do as They Please!' An Environmental History of a Finnish City in Wartime. *Environmental History, 9*(4), 679–700.

Malcolmson, R. W., & Mastoris, S. (1998). *The English Pig: A History.* London: Hambledon Press.

McNeur, C. (2011). The 'Swinish Multitude': Controversies over Hogs in Antebellum New York City. *Journal of Urban History, 37*(5), 639–660.

Ministry of Agriculture and Small Pig Keeper's Council. (1943). *Home Curing of Bacon and Hams – Bulletin No. 127 of the Ministry of Agriculture and Fisheries . . . Published in Conjunction with the Small Pig Keeper's Council.* London: Her Majesty's Stationery Office.

118   *Thomas Webb*

Ministry of Agriculture Memoranda. (1944). *Memorandum for Meeting at Kingston-upon-Thames, 16/12/1944*. Constitution of Pig Clubs (January 1940–March 1945) (MAF 84/55). National Archives, London, United Kingdom.

Moore, G. (1945). Scraps from the Kitchen Front. *Public Cleansing and Salvage, 35*(416), 308.

Otter, C. (2006). The Vital City: Public Analysis, Dairies and the Slaughterhouses in Nineteenth-Century Britain. *Cultural Geographies, 13*(4), 517–537.

Otter, C. (2008). Civilizing Slaughter: The Development of the British Public Abattoir, 1850–1910. In P. Young Lee (Ed.). *Meat, Modernity and the Rise of the Slaughterhouse* (pp. 89–106). Durham, NH: University of New Hampshire Press.

Pearson, C. (2015). Beyond 'Resistance': Rethinking Nonhuman Agency for a 'More-Than-Human' World. *European Review of History, 22*(5), 709–725.

Philo, C. (1995). Animals, Geography, and the City: Notes on Inclusions and Exclusions. *Environment and Planning D: Society and Space, 13*(6), 655–681.

Philo, C., & Wilbert, C. (2000). Animal Spaces, Beastly Places. In C. Philo & C. Wilbert (Eds.). *Animal Spaces, Beastly Places: New Geographies of Human-Animal Relations* (pp. 1–34). London: Routledge.

Pigs in the Ruins. (1943, April 8). *Picture Post*.

Riley, M. (2008). From Salvage to Recycling: New Agendas or Same Old Rubbish? *Area, 40*(1), 79–89.

Roodhouse, M. (2013). *Black Market Britain, 1939–1955*. Oxford: Oxford University Press.

Shelley, A.N.C. (1940, September 13). *Letter from A.N.C Shelley to M. Miller*. Poultry and Pig Keeping on Local Authorities' Housing Estates (1939–1951) (HLG 101/309). National Archives, London, United Kingdom.

Short, B., Watkins, C., & Martin, J. (Eds.). (2007). *The Front Line of Freedom: British Farming in the Second World War*. Exeter: British Agricultural History Society.

Small Pig Keepers Council. (1942). *SPKC Annual Report 1942*. Small Pig Keeper's Council – Annual Reports (1941–1950) (MAF 126/2). National Archives, London, United Kingdom.

Small Pig Keepers Council. (1945). *SPKC Annual Report 1945*. Small Pig Keeper's Council – Annual Reports (1941–1950) (MAF 126/2). National Archives, London, United Kingdom.

Small Pig Keepers Council. (1946). *SPKC Annual Report 1945*. Small Pig Keeper's Council – Annual Reports (1941–1950) (MAF 126/2). National Archives, London, United Kingdom.

Smith, V. (2007). *Clean: A History of Personal Hygiene and Purity*. Oxford: Oxford University Press.

Strang, J. W. (1939). Poultry Keeping in War Time. *Agriculture: The Journal of the Ministry of Agriculture, 46*(7), 660–663.

Thorsheim, P. (2013). Salvage and Destruction: The Recycling of Books and Manuscripts in Great Britain during the Second World War. *Contemporary European History, 22*(3), 431–452.

Thorsheim, P. (2015). *Waste into Weapons: Recycling in Britain during the Second World War*. Cambridge: Cambridge University Press.

Training to Save the Bacon. (1943, December 29). *Daily Mirror*.

Woods, A. (2012). Rethinking the History of Modern Agriculture: British Pig Production, c. 1910–65. *Twentieth Century British History, 23*(2), 161–195.

Worden, A. N. (1943). The Swill-Feeding of Pigs in Urban Areas. *Veterinary Record, 46*(55), 448–449.

Part III

The nation – historical animal bodies and human identities

9 Rebel elephants

Resistance through human–elephant partnerships

Jennifer Mateer

> The Enlightenment rational subject is not the only possibility for political subjec-
> tivity, but a historically-constructed convention that is laden with exclusions.
>
> – Nan Enstad (1998)

Historical accounts of political change generally focus on the impacts of human
policy and action on other populations of humans, past and present. The lives and
actions of animals enfolded in moments of political change, however, tend to be
obscured or reported as a "side note[s], not a central concern" (Hathaway, 2016,
p. 55). What opportunity is available to consider animals as political actors, particularly
in a historical context? This chapter seeks to address this question using the example
of elephants in South Asia. By examining the lives of elephants, and looking at how
they can be considered to have participated in shaping a shared human and nonhuman history, this chapter challenges an anthropocentric historical view of nature, the
environment, and the more-than-human world. The relationship between humans
and elephants has been and continues to be complex and dynamic, each possessing
agency, while also influencing the actions and development of the other.[1] Simply put,
beings other than humans have often changed the course of history.

The historical inclusion of elephants in many South Asian case studies is typically
concerned with the materiality of elephant bodies, counting them as tools of war,
beasts of burden, forms of capital, and objects for entertainment. In India, accounts
of elephant bodies of this nature are recorded as early as 6,000 BCE (Bist et al., 2001;
Choudhury, 1988), and as such, elephants have long been part of social, economic,
and political systems (Whatmore, 2002). However, even with the integration of
human and elephant lives, Jamie Lorimer (2010) has noted that "we know very little
about the materialities of interspecies relationships. Traces of human–elephant companionship must be gleaned from the margins of existing work" (p. 495).

Bringing to light accounts of elephants as political agents within the spaces they
share with humans is an important aspect of critical scholarship on nonhuman (or
more-than-human) actors, which challenges commonly understood concepts of animals as passive rather than active (Lorimer, 2010). This sub-discipline of more-than-
human geography argues that relegating animals, such as elephants, to the category
of "other" allows humans to ignore the impacts, fates, and political subjectivities of

animals (Collard, 2013; Hobson, 2007; Philo and Wilbert, 2000; Wolch and Emel, 1998). In South Asia – and India in particular – elephants can be seen to be effective political agents, and thus as part of a dynamic system that includes both human and more-than-human worlds and actors (Robbins, 2003; Hobson, 2007).

This chapter takes aim at the variety of ways that human–elephant relationships have played out in South Asia throughout time. First, I offer an analysis of the political subjectivity through the elephant's position within military systems and rebel activities. Next, I explore the ways that elephants are agents and objects of power and wealth in the region, often as a form of conspicuous consumption. Finally, I turn to the historical role elephants have served as "beasts of burden" for hauling and transportation. These three layers of elephant subjectivity inform the final portion of this chapter, which examines the role of political subjectivity of elephants in the context of rebellions and times of political instability. As such, this chapter demonstrates how elephants are subjects in politics and of political practices (Hobson, 2007).

Historical inclusion of elephants in the military

Beginning as early as 1100 BCE, elephants were key to military operations in South Asian nations (Csuti, 2006; Gadgil and Guha, 1993) – so much so that elephants are often referred to as the "tank of the ancient world" (Csuti, 2006, p. 17). Some historical accounts indicate that elephants were first domesticated specifically for these purposes (Bist et al., 2001). Accounts of elephants as military combatants exist in descriptions of many famous South Asian battles, including between Alexander the Great and King Porus (the ruler of Punjab) in 326 BCE and the Battle of Shakkar Khera in 1724 ACE (ibid.). During these and other military operations, elephants were "instruments of coercion . . . [just as] metal swords and shields" (Gadgil and Guha, 1993, p. 31).

Due to their importance in these and other military situations, a specific set of policies were developed in South Asia called the Arthasastra, which codified the appropriate methods to capture and care for elephants during and beyond times of war. Although many analyses of these policies may consider the Arthasastra a method of objectifying elephants, it can also be seen as part of a relational approach to political elephant agency. As suggested by Schurman and Franklin (2016), creating these types of relationships between humans and animals brings about a certain embodied knowledge, communication, and shared lived experiences, which opens up "spaces of everyday human–animal relations [which] become significant as settings for shared everyday life" (p. 42). This experience is made possible because, being responsible and accountable for the care of another being is not an "ethical abstraction; these mundane, prosaic things are the result of *having truck with each other*" (Haraway, 2008, p. 36, emphasis added). With this relational approach to understanding the human–elephant relationships, the elephants demonstrate their participation "in the production of the relationships" (Schurman and Franklin, 2016, p. 42).

Elephants lost their place in the military in the 1900s, when field artillery became common within South Asian militaries. However, even with their diminished roles on the battlefields, elephants were still part of military functions by supporting various bases as necessary parts of the supply chain (Csuti, 2006). For instance, during World War II, both Japanese and British forces employed elephants to manoeuvre through dense jungle terrain. This maneuverability became particularly important during the decisive battle of Mandalay in 1945, when British forces were able to utilize the partnership between soldier and elephant to capture Japanese soldiers, employing the unique physicality of these animals to move quickly into advantageous, but difficult, positions (Csuti, 2006). This gestures toward the relational agency and "truck" discussed by Haraway (2008, p. 36), where humans and nonhumans craft and co-create the worlds we inhabit together in provisional, asymmetrical, but often durable ways.

Elephants as historical symbols of wealth and power

Beyond considering elephants for their military potential, recent scholarship has also examined how elephants have functioned as symbols of power and wealth in India and other South Asian nations. To capture, train, and maintain an elephant population requires substantial financial resources. Thus, the display of elephants in public can be considered a symbol of surplus wealth and power, conferring social status to those who keep elephants (Csuti, 2006). Long true for Indian kings, this practice was also taken up by occupying European colonial officers.

Not only were elephants a mechanism by which to demonstrate power and wealth, but they were also used as a form of currency in South Asia. Kings often demanded elephant tributes from dependent or conquered states (Csuti, 2006). Alexander the Great also considered elephants a form of payment from the kings he conquered in order to become a member state of the empire. In fact, membership was "restricted to those who could supply at least one elephant to the state" (Sharma, 1991, p. 368), and those who could not afford the tribute of elephants to Alexander the Great were relegated to living as slaves or labourers. This is probably one of the most material examples of elephants being reduced to objects – in this case as a form of exchange – and yet their political value and position as actors cannot be denied. In this instance, elephants must be considered geopolitical actors, due to the way in which their presence, or absence, could render human individuals political subjects or slaves.

Elephants as a mobility assemblage

Perhaps one of the oldest practices enfolded in human–elephant partnerships has been the service of elephants in transportation. Elephants are documented working in service to human projects to log forests, haul materials, and participate in the construction of buildings as early as 200 BCE in South Asia (Csuti, 2006). Using elephants in this way makes sense given that they are large, strong, and intelligent and thus can be employed for many different forms of labor. Indeed, elephants became so integral to life in South Asia that little trace remains of their enmeshment with humans – narrating these routine and everyday relationships was overlooked by historians (Csuti, 2006). What is not routine, however, is that elephants were, in part, the reason for the development of certain towns and human populations. Elephants made the transport of necessary resources possible, such as cereal grains and timber, which could be stored and moved over long distances (Csuti, 2006; Gadgil and Guha, 1993). Without these resources, cities and empires, especially those in the mountainous regions, could not have been built or developed to the extent that they were. In a very real sense, then, elephants made urban India (Csuti, 2006; Nyman, 2016).

The transportation made available by elephants often represents an extraordinary and unique mobility assemblage. One recent example of the use of elephants for transportation took place when roads were destroyed or impassable following the 2004 Indian Ocean tsunami (Brooke, 2005). Under these extreme conditions, elephants, along with their cargo (including humans), were the only beings able to travel safely through the unstable terrain (Shell, 2015). This practice has historical antecedents. For instance, in 1969, rains in the state of Assam flooded out roads, making treks to and from tea plantations impossible without elephant power (Barkataki, 1969). Elephants, then, made possible humanitarian relief for planters who required food and supplies from outside the region (Shell, 2015). In the most literal way, elephants saved human lives.

There are significant advantages to employing elephants for transportation in these situations rather than other animals (like horses), or even human-made vehicles. Elephants, due to their size, are more capable than most animals of crossing rivers, which can be both deep and turbulent (Shell, 2015). Furthermore, elephants' feet are incredibly complex, able to expand and contract depending on the consistency of the ground (Ramsay and Henry, 2001). Unlike other animals or human-engineered vehicles, this complex physical attribute makes mobility on unstable ground more feasible. As discussed by Shell (2015), the benefit of elephant employment in treacherous terrain is the way in which they can provide people, specifically their mahouts,[2] "with a transformed geographic perspective on the spatial distinction between land and water" (p. 65). Thus, elephants make transportation in both physical states, and the in-between states, possible and/or easier.

Moreover, in conflict situations, elephants are also often the only viable transportation option due to governance breakdown and political dissent (Shell, 2015). For instance, roads may be impassable due to institutional failure and the subversion of dissidents. This has occurred on a number of occasions in South Asia. One example occurred during World War II, when elephants were employed to evacuate civilians in order to escape an incoming Japanese platoon (Haberstroh, 2003; Pankhurst, 2010). During this conflict in 1942, the Japanese gained control of the Burmese capital of Rangoon, where military personnel from Britain and India were stationed (ibid.). Civilians fled with these military personnel because of well-founded fears of capture and subsequent imprisonment in forced labor camps, which had been perpetrated by the Japanese across Malaysia, Burma, Indonesia, and the Philippines (Haberstroh, 2003). Although the trek was difficult, the Indian, British, and Burmese migrants were able to evade the Japanese and escape to India from Burma because "the [Japanese] enemy was blocked by monsoon-swollen rivers at the border" (Pankhurst, 2010, n.p.). The turbulent river channels were no deterrent to the evacuees who had partnered with elephants for their escape. Instead, the elephants increased their mobility, and ensured escape from the invading forces, by crossing the Chaukan Pass, which had become impassable to those without elephants (Martin, 2013; Pankhurst, 2010; Shell, 2015). This example speaks to the constructive relationship between elephants and humans, wherein elephants were essential to eluding Japanese occupation and potential abuses (Routledge, 1996).

This example of human–elephant partnership subverts the "conventional discourses of positing humans and nonhumans as separate," bringing new settings and possibilities for their relationship into being (Nyman, 2016, p. 77). Furthermore, because of the symbiotic nature of the relationship formed through transportation, this human–animal hybrid foregrounds the concept of "animaling." In their paper, Birk et al. (2004) describe "animaling" as a situation when the identity and subjectivity of both human and elephant are changed. The change then promotes "a view of the animal as having access to knowledge and agency" (Nyman, 2016, p. 76). Agency cannot be considered a human-only trait, and animaling, in particular, brings this into focus. Animaling also provides a critique of the cultural production of the human–animal divide, thereby shifting the perspective of animals and their political subjectivities (Birk et al., 2004; Hobson, 2007). The case of elephants demonstrates how animaling works to unsettle assumptions about who can deploy agency and in what ways.

Rebel elephants: historical resistance in India

Partnerships between humans and elephants can be particularly useful for individuals wishing to increase mobility beyond state-sanctioned activities, which has often included the use of elephants to enhance illegal activities, such as smuggling and logging.[3] However, evidence also suggests that elephants have often been used by people wishing to evade governing authorities and to rebel against certain political regimes – particularly when the soil resembles "a sea of black mud" (Malleson, 1896, p. 329). Examples of elephant-based mobility for rebels can be found in the post-colonial period in India up until the late twentieth century (Shell, 2015).[4] Guerilla fighting forces have historically succeeded in taking on larger, more established military forces by using elephants as a means of transportation, which some "might consider archaic" (Young, 2016, p. 238), but which has often brought about a highly successful outcome.

The Sepoy Mutiny, also known as the Indian Mutiny, was an anti-colonial rebellion staged against the British forces between 1857 and 1858 (David, 2003; Hibbert, 1988; Malleson, 1896). This mutiny was one of the first amalgamated anti-colonial rebellions in India, and an important demonstration of Indian solidarity (David, 2003; Hibbert, 1988). The mutiny began after British soldiers indiscriminately slaughtered civilians as a punishment for a rebellion of Indian soldiers (ibid.). This cruelty inspired further uprising, but rather than face certain death at the hands of retributive British soldiers, many rebels fled India for neighbouring nations or into secluded mountainous terrain. Two of the rebel groups that were most successful in evading British troops employed elephants to move subversively, changing course and avoiding governing authorities (Malleson, 1896; Shell, 2015). In cases like this, "it is the animals themselves who inject what might be termed their own agency into the scene, therefore transgressing, perhaps even resisting the human placements of them" (Philo and Wilbert, 2000, p. 14). A specific South Asian example of this during the Sepoy Mutiny occurred when the rebels crossed the flooded Chambal River by partnering with elephants. They escaped the British soldiers because the British, who had partnered with horses, were unable to move around the mud of the flooded river (Duberly, 1859; Shell, 2015). As such, the elephants created "their own 'other spaces', countering the proper places stipulated for them by humans" (Philo and Wilbert, 2000, p. 14), which in this case were the British.

Another rebel group was also able to evade the British by successfully partnering with elephants. This rebel group crossed the Rapti River at a particularly dangerous section to escape into neighbouring Nepal (Howitt, 1864; Ball, 1858). Although the British did try to follow the Indian rebels across the river, they became trapped by the monsoon-soaked terrain and were unable to pursue them to the border (Shell, 2015; Ball, 1858). Because elephants, due to their size and physical attributes, do not need to travel by road, guerrilla activities beyond state surveillance become possible and even successful. These elephant-enabled acts of rebellion make clear that the European occupying forces were unable to maintain their spaces of sovereignty. Instead, the British colonial state physically and symbolically shrank "once the monsoon rains began" (Scott, 2009, p. 61) and when rivers flooded. This shrinking allowed for a geographic expansion for rebels, and contraction for colonial powers, which needed to wait until the dry seasons to regain control. Elephants then were actors in the evasion of authorities, allowing for the subversion of, and resistance to, colonial regimes in India.

Elephants as political actors

The "animal question" has interested academics across various disciplines of the social sciences and humanities, providing an extensive and diverse range of academic texts and a renewed interest in critical animal geographies. The animal question itself, though, is not new, nor are animals new to politics. Animals have always been political, most explicitly in how they are used to describe, as their opposite, the standard "political subject: the human" (Collard, 2013, p. 227). What separates humans and animals using this rationale is that, unlike humans, animal bodies are killable and/or disposable (Haraway, 2008). In contemporary politics, attention to animal bodies thus focuses on them economically, in large part because of the way animals are excluded from legal rights and can, therefore, be commodified as an edible, wearable, viewable, or even symbolic product. Many possible analyses are lost with this limited interpretation, however. New avenues of debate can become visible and discursively possible if we consider animals not as passive, one-dimensional pawns within political and economic relations (Collard, 2013, p. 228).

By considering animals as active political subjects, able to change history and provide avenues of emancipation, we have an opportunity for understanding diverse forms of power, which exist in both institutions and practices.

The elephants who worked with the Indian rebels to resist British colonizers may not have possessed the *same* political agency as human citizens of India, as they could not take part in democratic decision-making processes, nor express *verbally* their preferences for certain policies or political practices. However, this does not preclude elephants from being political actors (Hobson, 2007). These elephants, along with other animals, are part of political life and can change how politics are enacted and challenged (Hathaway, 2016; Hobson, 2007). This speaks to the difference between "Politics" and "politics" (Flint, 2003), where "Politics" refers to national and international negotiations and arrangements, whereas "'politics' refers to the spaces, peoples, and practices that both challenge institutions through non-traditional political avenues . . . [including] the politics of the 'everyday'" (Hobson, 2007, p. 252). Who and what makes up "politics" is more diverse than the subjects with which "Politics" engages. This suggests the necessity of considering political subjects beyond the standard state-based understandings, and instead involving subaltern voices, including the contributions and lived experiences of more-than-human populations.

Animal and more-than-human entities have often been excluded from political geographies, as they are rarely considered political subjects. However, this situation does not need to remain. Considering the Asian elephants in the cases herein as only a resource frames the elephants as having no "individual character, knowledge, subjectivity or experience" (Tovey, 2003, p. 196). Through this frame, elephants, as well as other animals, exist only as the objects of politics – never as subjects, which limits the understanding of a complex reality which sees elephants as part of the (uneven) distribution of power within India.

By engaging with animal political subjectivities in this way and amending standard ontological approaches, my concern is not specifically around animal welfare or social/environmental justice. Rather, I aim to be part of conversations overcoming the human/wildlife divide, which sees agency and subjectivity as held only by "rational" (read: human) actors. Elephants, and other animals, are clearly part of lowercase "politics" (Flint, 2003). Lowercase politics is exemplified by the ways in which animals move beyond the codification of "object" to be used for political ends, to become subjects who can be objects *as well as* emancipatory agents (Hobson, 2007). For example, as stated previously, during the Sepoy Mutiny when rebels partnered with elephants, they were able to escape the British military forces and cross the Chambal and Rapti rivers. These actions demonstrate how elephants are agents, challenging the state through a non-traditional avenue as part of (lowercase p) politics (ibid.). Thus, by considering and analysing the activities of India's rebel elephants, these beings can be conceptualized as political actors and subjects of resistance.

Recently, scholarship within the sub-discipline of critical animal geographies has actively engaged with this shift in ontological approaches with significant success and impact. Scholars in this sub-field have remapped the border between human and the more-than-human to "explore the complex nexus of spatial relations between people and animals . . . [teasing] out the myriad economic, political, social and cultural pressures shaping these relations" (Philo and Wolch, 1998, p. 6). This scholarship stands to change the scope of political questions, including who or what can be considered political subjects (Hobson, 2007). In the cases of elephant actors in India, this approach allows scholars to see the elephants as political subjects during political instability and humanitarian crises, rather than as static political tools or objects, making elephants active resisters to certain political practices[5] (Robbins, 2003). This does not mean that animals are "mapped onto the pre-existing human world or are dumb actors in diverse polities" (Hobson, 2007, p. 257). Instead, animals become part of the construction of history and politics, which had previously been considered only as a rational human-only creation.

Conclusions

Beyond the role that they have played in Indian rebel activities, elephants have consistently been tied to both everyday politics (lowercase p) and the international relations of (uppercase) Politics (Flint, 2003). This has occurred through the circulation of animal bodies as part of conservation and capital accumulation, as well as the multi-scaled approach to understanding and mitigating elephant–human conflict. Beyond this, some animals have been able to act as political subjects, not through their anthropomorphization nor through an animal rights-based discourse, but rather through the notion that agency and political subjectivity is a part of more bodies than just the "rational and enlightened" human subject. As such, my discussion has argued for a different understanding of who or what is part of politics, and for a relational understanding of agency. The elephants who assisted in humanitarian relief efforts in South and Southeast Asia and who partnered in resistance efforts against colonial authorities in India exemplify how animal geographies can redraw the lines of inclusion and exclusion in order to make room for animals as political subjects. This reframes animals not simply as identities to be saved or managed, but rather as part of the everyday negotiations of (lowercase) politics and political encounters (Flint, 2003) – and as actors rather than as objects. By acknowledging this position of animals, the conceptualization of political processes and outcomes can also change (Collard, 2013; Hobson, 2007), ensuring that historical transformations are more thoroughly understood.

Notes

1 Although animal populations beyond elephants should be considered to have agency, it is particularly interesting that elephants have been under-theorized given their similarities to humans. For example, both elephants and humans maintain long-lasting social relationships (Hathaway, 2016; Poole and Moss, 2008; Wittemyer et al., 2005). Both populations can learn, retain, and transmit information and skills to their respective social groups and can act in strategic and coordinated ways (Locke, 2013; Poole and Moss, 2008). Biologically, both populations have brain compositions, with a high neuron density and a developed neocortex, which make possible complex social behaviours and the

ability to learn difficult skills (Locke, 2013). Finally, both populations also form such strong familial bonds that burying and mourning the dead, followed by periodic visits to the gravesites, are common (Douglas-Hamilton et al., 2006; Poole and Moss, 2008). Although I do not argue that the cognitive similarities listed here are what denote the agency of elephants, it is important to note that humans are considered political agents based on parallel attributes.

2 A mahout is anyone working with elephants, including riding elephants. The term is most widely used in South and Southeast Asia (Bradshaw, 2009).

3 Logging, including both illegal and legal logging, includes the use of elephants is South and Southeast Asia. This is in large part a way to generate income for mahouts, who bring their elephants to work "hoping to find a better way of life for both the elephants and themselves" (Food and Agriculture Organization of the United Nations, 2002, p. 36). Elephants are considered especially useful in the logging industry because they are able to physically push down trees and drag the logs out to the road for transport. They are considered an "ideal tool" (Food and Agriculture Organization of the United Nations, 2002, p. 52) for the job, since they are able to avoid damaging the forest ecosystem, which is often caused by machinery or other logging devices (Blake and Hedges, 2004).

4 Although not in India, the Viet Cong used elephants employed for mobility as a way to evade American troops during the Vietnam War (Olivier, 1993).

5 Further examples of this ontological shift, including animals as political subjects, can be found in the book *Animal Geographies*, edited by Wolch and Emel (1998), and *Animal Spaces, Beastly Places*, edited by Philo and Wilbert (2000). Both of these texts include chapters wherein animals become central agents in the creation of space and place through their actions and identities.

References

Ball, C. (1858). *The History of the Indian Mutiny: Giving a Detailed Account of the Sepoy Insurrection in India*. London, UK: London Print and Publishing Co.

Barkataki, S. (1969). *Assam*. New Delhi, India: National Book Trust.

Birke, L., Bryld, M., & Lykke, N. (2004). Animal Performances: An Exploration of Intersections between Feminist Science Studies and Studies of Human/Animal Relationships. *Feminist Theory*, 5(2), 167–183.

Bist, S. S., Cheeran, J. V., Choudhury, S., Barua, P., & Misra, M. K. (2001). Domesticated Asian Elephants in India: Country Report. *Ministry of Environment and Forests*. New Delhi: Government of India.

Blake, S., & Hedges, S. (2004). Sinking the Flagship: The Case of Forest Elephants in Asia and Africa. *Conservation Biology*, 18(5), 1191–1202.

Bradshaw, G. A. (2009). *Elephants on the Edge: What Animals Teach Us about Humanity*. New Haven, CT: Yale University Press.

Brooke, J. (January 7, 2005). Thais Use Heavy Equipment: Elephants Help Recover Bodies. *New York Times*. Retrieved from: www.nytimes.com/2005/01/07/world/worldspecial4/thais-use-heavy-equipment-elephants-help-recover-bodies.html?_r=0.

Choudhury, A. U. (1988, June 12). The Marauders of Meleng. *The Sentinel* (Guwahati, India).

Collard, R. C. (2013). Panda Politics. *The Canadian Geographer/Le Géographer Canadien*, 57(2), 226–232.

Csuti, B. (2006). Elephants in Captivity. In M. Fowler & S. Mikota (Eds.), *Biology, Medicine, and Surgery of Elephants* (pp. 17–22). Oxford: Blackwell.

David, S. (2003). *The Indian Mutiny: 1857*. New York, NY: Penguin Books.

Douglas-Hamilton, I., Bhalla, S., Wittemyer, G., & Vollrath, F. (2006). Behavioural Reactions of Elephants towards a Dying and Deceased Matriarch. *Applied Animal Behaviour Science*, 100, 87–102.

Duberly, F. I. (1859). *Campaigning Experiences in Rajpootana and Central India during the Suppression of the Mutiny, 1857–1858*. London, UK: Smith, Elder and Co.

Enstad, N. (1998). Fashioning Political Identities. *American Quarterly*, 50, 745–782.

Flint, C. (2003). Dying for a 'P'? Some Questions Facing Contemporary Political Geography. *Political Geography*, 22, 617–620.

Food and Agriculture Organization of the United Nations. (2002). *Giants on Our Hands*. Bangkok, Thailand: Regional Office for Asia and the Pacific.

Gadgil, M., & Guha, R. (1993). *This Fissured Land: An Ecological History of India*. Berkeley: University of California Press.

Haberstroh, J. (2003). In Re World War II Era Japanese Forced Labor Litigation and Obstacles to International Human Rights Claims in U.S. Courts. *Asian American Law Journal*, 10(2), 253–294.

Haraway, D. (2008). *When Species Meet*. Minneapolis, MN: University of Minnesota Press.

Hathaway, M. J. (2016). Animals as Historical Actors? Southwest China's Wild Elephants and Coming to Know the Worlds They Shape. In J. Thorpe, S. Rutherford, & A. Sandberg (Eds.), *Methodological Challenged in Nature-Culture and Environmental History Research* (pp. 55–65). New York: Routledge.

Hibbert, C. (1988). *The Great Mutiny: India 1857*. London, UK: Allen Lane.

Hobson, K. (2007). Political Animals? On Animals as Subjects in An Enlarged Political Geography. *Political Geography*, 26, 250–267.

Howitt, W. (1864). *Cassell's Illustrated History of England*, vol. 8. London, UK: Cassell, Petter and Galpin.

Locke, P. (2013). Food, Ritual and Interspecies Intimacy in the Chitwan Elephant Stables: A Photo Essay. *The South Asianist*, 2(2), 71–86.

Lorimer, J. (2010). Elephants as Companion Species: The Lively Biogeographies of Asian Elephant Conservation in Sri Lanka. *Transactions of the Institute of British Geographers*, 35(4), 491–506.

Malleson, G. (1896). *History of the Indian Mutiny: Commencing from the Close of the Second Volume of Sir John Kaye's History of the Sepoy War*, vol. 3. London, UK: Longmans, Green and Co.

Martin, A. (2013). *Flight by Elephant: The Untold Story of World War II's Most Daring Jungle Rescue*. London, UK: Fourth Estate.

Nyman, J. (2016). Re-Reading Sentimentalism in Anna Sewell's Black Beauty: Affect, Performativity, and Hybrid Spaces. In J. Nyman & N. Schuurman (Eds.), *Affect, Space and Animals* (pp. 65–79). New York: Routledge.

Olivier, R. (1993). Distribution and Status of the Asian Elephant. *Oryx*, 14, 379–424.

Pankhurst, N. (November 1, 2010). 'Elephant Man' Who Staged Daring WWII Rescues. *BBC News*. Retrieved from: www.bbc.com/news/uk-11652782.

Philo, C., & Wilbert, C. (Eds.). (2000). *Animal Spaces, Beastly Places: New Geographies of Human-Animal Relations*. London, UK: Routledge.

Philo, C., & Wolch, J. (1998). Through the Geographical Looking Glass: Space, Place, and Society-Animals Relations. *Society and Animals: Journal of Human-Animal Studies*, 6(2), 103–118.

Poole, J. H., & Moss, C. J. (2008). Elephant Sociality and Complexity: The Scientific Evidence. In C. Wemmer & C. A. Christen (Eds.), *Elephants and Ethics: towards a Morality of Coexistence*. Baltimore: The John Hopkins University Press.

Ramsay, E. C., & Henry, R. W. (2001). Anatomy of the Elephant Foot. In B. Csuti, S. Bechert, & U. S. Bechert (Eds.), *The Elephant's Foot* (pp. 9–12). Ames, IA: Iowa State University Press.

Robbins, P. (2003). Political Ecology in Political Geography. *Political Geography*, 22(6), 641–645.

Routledge, P. (1996). Critical Geopolitics and Terrains of Resistance. *Political Geography*, 15, 509–531.

Schurman, N., & Franklin, A. (2016). In Pursuit of Meaningful Human Horse Relations: Responsible Horse Ownership in a Leisure Context. In J. Nyman & N. Schuurman (Eds.), *Affect, Space and Animals* (pp. 40–51). New York: Routledge.

Scott, J. (2009). *The Art of Not Being Governed: An Anarchist History of Upland Southeast Asia*. New Haven, CT: Yale University Press.

Sharma, R. S. (1991). *Aspects of Political Ideas and Institutions in Ancient India* (3rd ed.). Delhi: Motilal Banarsidass.

Shell, J. (2015). When Roads Cannot Be Used: The Use of Trained Elephants for Emergency Logistics, Off-Road Conveyance, and Political Revolt in South and Southeast Asia. *Transfers*, 5(2), 62–80.

Tovey, H. (2003). Theorising Nature and Society in Sociology: The Invisibility of Animals. *Sociologia Ruralis*, 43, 196–215.

Whatmore, S. (2002). *Hybrid Geographies: Natures Cultures Spaces*. London: Sage.

Wittemyer, G., Douglas-Hamilton, I., & Getz, W. M. (2005). The Socioecology of Elephants: Analysis of the Processes Creating Multitired Social Structures. *Animal Behavior*, 69, 1357–1371.

Wolch, J. R., & Emel, J. (Eds.). (1998). *Animal Geographies: Place, Politics, and Identity in the Nature-Culture Borderlands*. London: Verso.

Young, R. R. (2016). Review: Transportation and Revolt. *Transportation Journal*, 55(2), 237–239.

10 Western horizons, animal becomings

Race, species, and the troubled boundaries of the human in the era of American expansionism

Dominik Ohrem

"This country of strange metamorphose"

What is the American West? And where? These deceptively straightforward questions have accompanied, and often troubled, those generations of scholars that have grappled with the idea and history of the "West" (as well as its conceptual partner in crime, the "frontier"), from its gradual cultural emergence in the early 1800s to the Manifest Destiny–fueled 1840s to what historian David Wrobel (1993) has termed the "postfrontier anxiety" of the Progressive Era.[1] Specifying the *where* of the West has been complicated by longstanding disagreements among historians about its precise boundaries and the criteria on which they should be based (cf. Nugent, 1992). The *what* of the West comprises at least two subquestions: the first is whether the West should be conceived of in terms of a place or region or whether it can be understood more adequately as a process, in line with the processual character of westward expansion and the shifting, ephemeral sociospatial arrangements associated with it (cf. Limerick, 1987; R. White, 1991). This latter idea resonates particularly well with those liminal (time-)spaces referred to as "frontiers," which accompanied this process and its ruthless uprooting of peoples and species, cultures and ecologies. The second subquestion, on the other hand, concerns the geographical and historical reality of the West as such, and the extent to which this reality has been co-shaped by mythology, fantasy, and desire – perhaps to the point of indistinguishability (cf. H. N. Smith, 1970; Slotkin, 1985; Grossman, 1994). In other words, the reason the West has always been more and less than a reference to a range of particular landscapes or topographies – the Great Plains, the Rocky Mountains, the Great Basin, the Sierra Nevada, and so on – is because its existence as a distinct physical geography ultimately remains inseparable from, and frequently finds itself in tension with, its existence as an *imaginative* geography. This characteristic elusiveness of the West was not lost to more skeptical contemporaries, some of whom questioned the increasing number of sensationalist accounts of the promises or dangers of the West – even as they themselves, in their own work, sometimes contributed to this blurring of the line between Western reality and fantasy. As the traveler and famous painter, George Catlin, remarks about the idea of the West, "[f]ew people even know the true definition of the term" or the location, boundaries, and nature of this "phantom-like" country, "whose fascinations spread a charm over the mind almost dangerous to civilized pursuits" (1841, p. 62).

One of the major developments of the nineteenth century, the process of territorial expansion into the regions west of the Mississippi had already been anticipated by the purchase of the enormous Louisiana Territory in 1803 and a number of scientific-military expeditions, such as the Lewis and Clark expedition from 1804–06. Even though territorial expansion remained a contested issue in American society – not least due to its connection with the possible extension of chattel slavery into Western states and territories – for Thomas Jefferson and other early proponents of exploration and expansion Western geographies represented a vast domain of untold economic, political, and cultural potential. From the onset of territorial expansion, the West and the frontier have often been associated with notions of transformation or metamorphosis, of becoming-other. As historian Richard Slotkin argues in his formative frontier trilogy, a widely held belief in the frontier's "capacity to work grand transformations on the character, fortunes, and institutions of the inhabitants" accompanied its movement across the continent, and while discourses and imaginings shifted with the different stages and contexts of westward expansion, they all "shared a common implication that transformation would be part of the experience" (1985, p. 40) – an assessment echoed by Catlin's remark that "[t]he *sun* and *rats* alone . . . could be recognised in this country of strange metamorphose" (1841, p. 60, emphasis in original).

In this chapter, I adopt an angle on the history of American westward expansion that so far has not garnered sustained critical attention: the role of the West – understood here not so much as a more or less clearly bounded region but as a natural-cultural phenomenon that involves a dynamic ensemble of discourses, imaginings, sociospatial arrangements, corporeal experiences, and geo- and topographical materialities – for contemporary ideas about the human, the animal, and the boundaries between them. I argue that during a time in which emerging conceptions of humans' evolutionary kinship with nonhuman life and post-Enlightenment developments in fields such as comparative anatomy and natural history had already begun to make what it meant to be human less certain than ever before, the West (also) functioned as a domain of ontological speculation about, and experimentation with, the increasingly troubled boundaries of the human.

Such anxieties about the precariousness of definitions of humanity in contradistinction to a domain of "inferior" animality that could be less and less kept at a safe ontological distance cannot be understood in isolation from the broader epistemic shifts in European and settler colonial societies in the lead-up to the "Darwinian revolution." Nor was what I have elsewhere referred to as the "zooanthropological imaginary"[2] (Ohrem, 2017) solely shaped by the phenomenon of westward expansion. Yet for many antebellum Americans, the lifeways of and relations between the inhabitants of Western environments – both human and nonhuman, white and non-white – appeared to offer privileged (if also troubling) insights into the broader differences between humans and other species, the implications of human animality, and the supposed (racial) hierarchies among different groups of human beings. Such concerns were also reflected in the prominent role played by the "new" Western environments manifestly destined to become a part of the American republic for contemporary definitions of "savagery" and "civilization,"[3] a dichotomy that was tied to a racialized segmentation of humanity into the figure of white, civilized Man and what cultural theorist Sylvia Wynter calls "Man's human Others" (Scott with Wynter, 2000, p. 174; also see Wynter, 2003). The dynamics of race and species also informed what many contemporaries perceived as the dark underside of the West's transformative potentials: its supposed tendency to produce liminal forms of "humanimality" that not only seemed to elude, and thus challenged, hegemonic categorizations of humanity and animality but, in their ontological indeterminacy, also threatened to dissolve the boundary between the figure of Man and its (Indigenous) human Others.

"Animality," writes philosopher Dominique Lestel, "remains a horizon of the human, that of its loss or escape outside of itself" (2014, p. 62). Lestel's remark serves as a reminder that the animal nature of the human continues to be something of a conceptual problem for a cultural and historical tradition that has long refused, or struggled, to think of the human "in the texture of animality" (p. 64). As a "horizon," animality seems to be something distant yet never out of view, something one can move or be drawn toward, something that can be both threat and promise, "loss" and "escape." From the perspective of many nineteenth-century Euro-Americans, Western horizons were coextensive with horizons of animality, of becoming animal or unbecoming human, which means that in the context of westward expansion there was a distinctly spatial element to both "animality-as-loss" and "animality-as-escape." In this way, nineteenth-century conceptions of the West often invoked both a geographical and an ontological horizon.

The lure of Western horizons

While the history and mythology of American westward expansion is mostly associated with the movement into the trans-Mississippi regions beginning in the antebellum period, to some extent concerns about the prospects and risks of westward expansion predate the intensification of expansionism in the first half of the 1800s. This is particularly true with regard to eighteenth-century American ideas about wilderness and frontier life and the physical and moral degeneracy associated with the figure of the frontiers- or "backwoodsman." For example, in the 1744 travel journal of Maryland physician Alexander Hamilton, he informs a young man who asks him about the potential dangers lurking in the "terrible woods" on the way to Maryland that "the most dangerous wild beasts in these woods were shaped exactly like men, and they went by the name of Buckskins, or Bucks, tho' they were not Bucks either, but something, as it were, betwixt a man and a beast" (1907, p. 150; also see Lemay, 1978). In this early, pejorative usage in Hamilton's *Itinerarium*, the term "buckskin" evokes the idea of a liminal being, a beast-like human or human-like beast, that inhabits the wilderness environments beyond the pale of civilized society, law, and government, an image that ties in with the traditional European figure of the Wild Man. Prominently epitomized by the likes of George Armstrong Custer and William "Buffalo Bill" Cody, in the second half of the nineteenth century, buckskin clothing would quickly become a potent symbol of rugged, masculine Americanness. But for Hamilton, the wild animal skins that enveloped (formerly) civilized bodies signified something else: the possibility that civilized whites' exposure to uncivilized environments and lifeways produced forms of humanity that constituted a threat, not only to the life of individual Americans but also to a broader social order. Given his status as a member of the gentry, no doubt class plays a certain role in Hamilton's description. But the image of violent, beastly semi-humans roaming through the woods also evokes what Hayden White in his discussion of the Wild Man tradition refers to as the specter of "species corruption" (1972, pp. 9, 14–15). This phenomenon not only pointed to the fragility of the social bonds of civilized humanity, but also called into question the very concept of civilized Man as that type of humanity defined by its ontological distance from both nonhuman animals and animal-like humans.

More directly anticipating the ways in which anxieties about species corruption and social order became tied to the phenomenon of westward expansion, French American writer J. Hector St. John de Crèvecoeur addresses the supposedly degenerative, "animalizing" tendencies of frontier life in his semi-autobiographical *Letters from an American Farmer* (1782). In the famous third letter, "What Is an American?," Farmer James, the fictional narrator of the *Letters*, laments that in the wild regions farthest from the civilized society of the Atlantic seaboard, "men appear to be no better than carnivorous animals of a superior rank," with their degenerate inhabitants existing in "a perfect state of war" both with the wild beasts always prowling in the vicinity and the increasingly beast-like members of their own species (1904, p. 59). James explains:

> By living in or near the woods, their actions are regulated by the wildness of the neighbourhood. The deer often come to eat their grain, the wolves to destroy their sheep, the bears to kill their hogs, the foxes to catch their poultry. This surrounding hostility, immediately puts the gun into their hands.
>
> (p. 66)

Crèvecoeur's environmental determinism and the widespread idea of the contagious wildness inherent to "undomesticated" environments beyond the corrective influence of civilized human sociality is embodied here by forms of wild animal agency and the human responses provoked by them – a fateful reciprocity through which the white men of the frontier are ultimately as much claimed by their environment as the supposedly inferior animals and savage humans around them, making them "ferocious, gloomy, and unsociable" in the process (p. 67).

Crèvecoeur's portrayal of frontier life in terms of a violent and disorderly Hobbesianism that pits civilized creatures – white humans and their domesticated animals – and a fledgling civilization against a seemingly omnivorous wilderness, starkly contrasts with the ideal of trans-species patriarchy depicted in the preceding letter on the "Situation, Feelings, and Pleasures of an American Farmer." Here, Crèvecoeur paints the picture of a peaceful rural setting in which the benevolent rule of the white farmer-patriarch extends not only to his wife, children, and, at least potentially, enslaved black servants, but also to the nonhuman creatures on his farm. As James tells us, attending to his cattle after a severe winter "is one of those duties which is sweetened with the most rational satisfaction," and he amuses himself by "beholding their different tempers, actions, and the various effects of their instinct now powerfully impelled by the force of hunger" (p. 32). Drawing an analogy between his own hierarchical relationship with the subordinate creatures on his farm and the willing subordination of civilized human citizens to rational republican government, he explains that,

> the law is to us precisely what I am in my barn yard, a bridle and check to prevent the strong and greedy, from oppressing the timid and weak. . . . Some I chide, others, unmindful of my admonitions, receive some blows. Could victuals thus be given to men without the assistance of any language, I am sure they would not behave better to one another, nor more philosophically than my cattle do. . . . Thus by superior knowledge I govern all my cattle as wise men are obliged to govern fools and the ignorant.
>
> (pp. 32–33)

The farmer-patriarch's pastoral domain is defined by the permeable yet strictly policed spatial boundaries of the farm and the ways in which nondomesticated creatures are allowed into, or are repelled from, this domestic(ated) environment. Thus, James writes about the "great fund of pleasure" he draws from his interactions with quails and from the "agreeable spectacle" of the "beautiful birds, tamed by hunger, intermingling with all my cattle and sheep" (p. 31). In contrast to the corruptive tendencies endemic to life in frontier environments, this intermingling of "tame" and "wild" animate forms takes place within the safe confines of the domestic order governed by the farmer, who decides about the kinds of species as well as the forms and intensities of interspecies relations permitted in his domain in accordance with a broader, civilized socionatural order.

Ideas about the environmental malleability of humans' physical and mental constitution were not limited to the transformative agency exerted by nonhuman natural forces and landscapes (cf. Valencius, 2004; L. Nash, 2006); they also included the bodily and social *practices* and *relations* – including those between humans and other species – that supposedly characterized life in a particular location. As the enduring fascination with the life of the frontiersman shows, however elusive conceptions of the West may have been in terms of actual physical geography, there has been a strong continuity in how the sociospatial arrangements of Western life have been characterized with reference to an ensemble of practices and relations that differed substantially from, and were in many ways (seen as) antithetical to, the established norms and parameters of civilized life in the rural and urban East.

During a crucial transitional period that saw a gradual shift in conceptions of race from a focus on its cultural–environmental malleability to its rearticulation in biological essentialist terms, anxieties about species corruption became increasingly interwoven with anxieties about *racial* corruption, an ominous twin specter haunting the always precarious whiteness and humanness of Man. As Paul Outka writes in *Race and Nature*, if the transformation of savage wilderness into idyllic and productive pastoral landscapes "was the privileged sign of the white settlers' racial superiority" and functioned "simultaneously, and contradictorily, as the origin of whiteness and the result of it," Crèvecoeur's description of frontier life testifies to the seeming "fluidity between natural and racial identity" as well as wild nature's "alchemic power over racial identity" (2008, p. 32). It might be worthwhile, however, to pay more specific attention to the role of animality in this context instead of (implicitly) subsuming this aspect under a generic concept of nonhuman nature. This is because Crèvecoeur's contrasting descriptions of the white farmer's pastoral order and the race–species corruptions associated with Western frontier environments hinge not only on the general opposition of wild and domesticated nature, but on a more specific distinction between savage and civilized human–animal relations and between those corporeal and social practices conforming to a normative concept of civilized humanity and those expressive of a semi-human animality frequently projected onto the bodies of racialized others.

The human animal and/in the trans-Mississippi West

Crèvecoeur's "West" was still located within the bounds of the familiar Eastern woodlands. However, in the early decades of the nineteenth century, imaginings of the West had already begun to shift their focus to the regions beyond the Mississippi, which differed markedly from the forested East and came with unique challenges, beginning with the very attempt at aesthetic description (cf. Hyde, 1990). "Wilderness" was a well-established concept widely employed to characterize these bioregions of the trans-Mississippi West, which – like the "Great American Desert," as explorer Stephen H. Long famously termed the western part of the Great Plains – seemingly "resisted both agrarian settlement and white bodies" (LeMenager, 2004, p. 16). On the one hand, the conceptual substance of wilderness has always been less about geo- or topographical specifics than about the contrasting juxtaposition of civilized geographies with those imagined as yet beyond the transformative reach of white civilization (cf. Cronon, 1995; Flores, 1999; R. F. Nash, 2014). But as much as wilderness has always been a Euro-American construct and an element of hegemonic discourses that underpinned a racialized ontology of Man, cultural imaginaries of wilderness also evoked forms of environmental and animal agency different from those normally encountered – and accepted – in the more anthropogenic and anthropocentric rural and built environments east of the Mississippi, with their strongly regulated forms and spaces of human–animal encounter and interaction. When antebellum travelers remarked on the "wild liberty" of the mustang (Anderson, 1967, p. 167) or the prodigious size and thundering migrations of bison herds, they acknowledged the relative autonomy of animal bodies and movements in expressly more-than-human environments that seemed to challenge Manifest Destiny's fantasies of the supremacy of American Man while also providing the necessary setting for narratives and performances of Western "conquest."

According to the dominant story of American westward expansion, which found its most (in)famous expression in Frederick Jackson Turner's 1893 "frontier thesis," the experience of the West promoted the emergence of a specifically American type of racial/national and masculine identity embodied by the ruggedly individualist and fiercely democratic "Anglo-Saxon" frontiersman. For Turner and other *fin-de-siècle* commentators, such as Theodore Roosevelt or Frederic Remington, struggling against a continental wilderness and its savage human and animal inhabitants was equivalent to the very process of becoming American, with the ever-shifting frontier functioning as "the line of most rapid and effective Americanization" (Turner, 1961 [1893], p. 39). The immersion of civilized Americans into Western wilderness environments was thus always clearly prescribed in terms of its ultimate result or "product" – American manhood and, by extension, the very quality of "Americanness" as such. If, as Turner has it, on the frontier the "wilderness master[ed] the colonist" (p. 39), this "mastery" was only ever temporary and partial. And if each stage of frontier expansion meant a "return to primitive conditions" (p. 38), what the frontiersman actually and eventually "returned" to was a uniquely American manifestation of civilized whiteness and superior humanity.

In contrast to such *fin-de-siècle* celebrations of frontier masculinity, for many antebellum Americans the extent to which the transformative agency of the West could indeed be understood as congenial with the master narrative of civilizational progress was an open question rather than a foregone conclusion. Besides the savage lifeways of Western Indigenous people(s), it was in particular the markedly ambivalent rather than unequivocally "heroic" figure of the frontiersman that captured Americans' attention. As early Western writer James Hall explains in his *Letters from the West* (1828), the idea of westward "emigration" used to "carr[y] with it many unpleasant sensations," invoking the picture of "a respectable man hieing to an unknown land, to seek a precarious existence among bears and musquitoes [sic]" (p. 171). The persistence of this undercurrent of anxiety regarding the effects of Western environments on individual Americans and American society more broadly is also evident in an 1833 article published in Hall's short-lived *Western Monthly Magazine*, which repudiates the Crèvecoeurian image of the frontiersman as "a solitary, unsocial being, living separate from his species" (1833, p. 52). While Hall concedes that this may indeed be "the history of some; and at certain periods of their lives, of many," he laments Easterners' seeming inability to recognize the "manliness of the western character," insisting that, on the whole, Westerners "are far from being an unsocial race" and a "people equally removed from the savage ferocity of the wild Irishman, and the sullen stupidity of the English peasant" (p. 49). For Hall, the idea that Westerners were inferior to the inhabitants of other parts of the country was not merely wrong but in fact "unphilosophical," because, while geographical factors such as climate "may affect the human skull, advantageously or otherwise," according to Hall's racialized proto-anthropology, this aspect was of secondary importance compared to the fact that "men's brains – at all events, the brains of white men – are made alike, all the world over" (p. 50).

In Hall's time, imaginings of the West were already becoming a staple of a burgeoning print culture of books and periodicals, with more and more narratives of Western travel and adventure circulating among an ever-wider readership. For many antebellum whites like journalist and adventurer Charles Wilkins Webber, Western wilderness environments and the "extraordinary relations" they established between "the civilized man, the savage and the brute" (1851, p. 30) offered fertile grounds for proto-Darwinian conceptions of human–animal difference and interspecies relations, especially those between humans and other creatures. Fittingly enough, Webber's own biography in many ways epitomizes the brashness of antebellum expansionist endeavors. Moving from his home state of Kentucky to the conflict-ridden Texas frontier of the late 1830s to accompany a group of Texas Rangers, he later organized an ill-fated expedition to Arizona Territory, attempted to improve transportation in the Southwest with a "camel company," and eventually died in Nicaragua while participating in William Walker's infamous mid-1850s filibustering exploits.[4] In 1844, following his time in Texas, Webber moved to New York, where he worked as a journalist and as editor for the *American Whig Review*, publishing several articles and stories about his wilderness adventures, the "erratic wanderings" which had brought him into familiarity "with all wild, grotesque and lonely creatures that populate those infinite solitudes of nature" (p. 90). According to Webber's red-in-tooth-and-claw conception of animal life, the principal form of interspecies relationship was that between hunter and hunted, with humans figuring as apex predators in a universe defined by a relentless struggle for existence in which "the *strong*, of course, conquer" (p. 21, emphasis in original). But even if humans were taken out of the equation, Webber's naturalization of violence as the defining characteristic of both intra- and interspecies relations was supported by the workings of the animal world itself; for proof one needed to only look at the "lustful battles of the animal tribes among themselves" (p. 66). After briefly referring to the "savage contests of the canines, felines, &c," Webber elaborates on his point by drawing on his experiences with animal life on the "great prairies" of the West:

> It is a fact, with regard to the habits of the Mustangs . . . that the weaker stallions are invariably, after desperate contests, either killed or driven into solitary banishment, from which they never return to the herd, until their strength and prowess have been so far developed in the solitude, as to give them some hopes of being able to triumph in a renewed struggle with their conquerors.
>
> (p. 22)

Francis Parkman's account of his 1846 travels on the Oregon Trail similarly emphasizes the violent existential struggle that seemingly permeated animal life in the West in a way that also suggested an interpretation of American expansionism as a teleologically preordained "interspecies" struggle between civilized whites and savage indigenes. A wealthy Bostonian plagued by a debilitating neurological disease, Parkman attempted to overcome his condition by immersing himself into the invigorating harshness of the Western wilderness, an environment that forcibly reduced "the human biped . . . to his [sic] primitive condition" (1996, p. 252) and threatened to make "quick and sharp work" (p. 183) of all those too weak to adapt to and cope with its unforgiving realities. More virulently than Webber, Parkman points to the ways in which Western environments functioned as imaginary–experiential spaces from which conceptions of civilized Man and its human and nonhuman Others emerged through the discursive dynamics of race and species. For Parkman,

> a civilized white man can discover but very few points of sympathy between his own nature and that of an Indian. . . . [A]n impassable gulf lies between him and his red brethren of the prairie. Nay, so alien to himself do they appear, that having breathed for a few months or a few weeks the air of this region, he begins to look upon them as a troublesome and dangerous species of wild beast.
>
> (p. 237)

While 'brethren' seems to locate Indigenous people within the sphere of human kinship, the notion of an 'impassable gulf' is strongly evocative of the unbridgeable divide that supposedly separates humans from other living beings, a conceptual move, which reframes ideas about racial differences among humans in more radical terms of interspecies difference. Indeed, for Parkman, rather than revealing the ties of human kinship, breathing the "air of this region" for some time only served to further highlight the inferior otherness of Western indigenes and their "naturally" antagonistic relationship with civilized Man. During his own travels in the Southwest, Parkman's contemporary Frederick Law Olmsted echoes this violently fatalistic sentiment. Writing about the quasi-genocidal attitudes, widespread among the white "borderers" that relegated Southwestern Native Americans to the status of "blood-thirsty vermin" in need of extermination, Olmsted recounts how this sentiment also shaped his own encounters with them: "A look into their treacherous eyes," he remarks, "was enough to . . . rouse the self-preservative tigerhood of the animal man, latent since we ran naked like the rest in the jungles" (1857, p. 297).

Such dehumanizing portrayals of Indigenous people(s) expressed in antebellum writing about the West should not distract us from the ambiguity surrounding the interplay of race and species in this context. Olmsted's remarks on the "disgusting brutishness" of Native Americans and on the "excellent suggestion" that "their young, like those of other animals, can be caught and tamed" as a more "humane" alternative to their extermination (pp. 297–298) relegates them to an animal-like status in radical contradistinction to the *humanitas* of civilized Man. But Olmsted's image of a primordial human "tigerhood" also suggests that, rather than being specific to the brutish "nature" of Western indigenes, the savagery that suffused Western environments merely brought to the surface those predatory qualities of "the animal man" that civilization had been able to keep in check, but never fully transcend, through the "artifices" of civilized society. As Washington Irving concludes from his 1832 travels into Indian Territory: "Man is naturally an animal of prey; and, however changed by civilization, will readily relapse into his instinct for destruction," with his "ravenous and sanguinary propensities daily growing stronger upon the prairies" (1835, p. 125).

Viewed from this angle, the Western experience highlighted Man's status as a precarious civilizational achievement, a being whose existence was tied to the sustaining framework and corrective influences of civilized sociality. Animality thus seems to be uneasily positioned within a field of tension between two discursive trajectories: on the one hand, the "naturalistic . . . conception of human nature" (Curti, 1980, p. 187) that took shape in the antebellum era meant that animality was increasingly seen as infusing even the vaunted figure of Man, posing a challenge to the ways in which both human exceptionalism and white civilization were conceived of in terms of non-animality. On the other hand, the well-established discursive strategy of disassociating Man from the taint of animality and displacing the latter onto the racialized domain of Man's human Others remained integral to legitimizations of both the "peculiar institution" of Black slavery and expansionist violence against Indigenous societies.

This displacement of animality onto nonwhite bodies was nonetheless complicated by the fact that neither whiteness nor *humanitas* could be safely contained within the conceptual and geographical domain of civilization, a problem that was prominently embodied by the figure of the white frontiersman. Echoing earlier Crèvecoeurian environmental determinism, George Frederick Ruxton's *Adventures in Mexico and the Rocky Mountains* describes the "mountain men" he encounters during his travels as a "'genus' more approximating to the primitive savage than perhaps any other class of civilized man," as creatures whose "habits and character assume a most singular cast of simplicity mingled with ferocity, appearing to take their colouring from the scenes and objects which surround them," and while these men "may have good qualities, . . . they are those of the animal" (1847, pp. 241–242). Walking the oft-invoked line between savagery and civilization, frontiersmen occupied a precarious dual role: widely recognized as authorities of knowledge about Western life, with their altered bodies, peculiar habits, and troubling associations with "inferior" types of humanity, these men also became anxiously scrutinized objects of concern in the contemporary zooanthropological imaginary. What to make of those individuals who left behind – turned their back on? – civilization in favor of the vagaries of the wilderness? What if, contra Turner, instead of becoming properly American, westering Europeans in fact became animal and savage, barely distinguishable from the creatures that inhabited the wilderness they were supposed to conquer? At least in Ruxton's rather ambiguous view, these conquerors of a brute wilderness were themselves "just what uncivilised white man might be supposed to be in a brute state" (p. 242).

Ruxton's account is typical of the vagueness of nineteenth-century distinctions between the conceptual dualisms of animality and humanity (which informed zooanthropological discourse) and savagery and civilization (which informed racial discourse). As ontological categories, animal and savage, human and civilized frequently bled into each other, and while this permeability and malleability goes some way to explain the discursive efficacy of these categories, the ontological zones of indistinction opened up by this ambiguity arguably undermined as much as they reinforced the sovereignty of (American) Man. In this light, we can also understand contemporary concerns about the frontiersman as indicative of a struggle to erase, or at least domesticate, the troubling ambiguity of this figure – a process that is clearly discernible in Turner's apotheosis of frontiersmen as catalysts of American expansionism and nation building but also in their earlier condemnation as beastly, ferocious degenerates who endangered the integrity and social cohesion of the fledgling republic.

While the frontiersman continued to be regarded as a moral and social problem well into the nineteenth century, perceptions began to shift over the course of the antebellum era, as this figure was not only met with increasing popular curiosity but also turned into an object of philosophical and scientific reflection at the intersections of zooanthropological and racial discourse. It may thus not be all too surprising that Ruxton's mountain men also make an appearance as "evidence" in a multi-piece article printed in the British natural history journal *The Zoologist*, which deals with the much-contested boundary between (human) reason and (animal) instinct and features a discussion of forms and stages of "retrograded" humanity. Compared to Western Indigenous peoples like the Paiute and Shoshone, who represented a type of humanity that had descended to the level of "the human creature," Ruxton's mountain men serve as an intermediate example indicating the "earlier stages" of this retrogressive process (Atkinson, 1859, p. 6315). In this liminal stage, the author explains, civilized whites had already undergone significant transformations that went beyond superficial behavioral habits and involved the adaptation of their sensory apparatus to the requirements of savage life – such as the "peculiar keenness of vision and intuitive perception of locality and direction" – necessary to survive in the wilderness, while, at the same time, "the higher portions of man's intellectual nature are as little apparently employed as possessed" (p. 6317). Did this transformative process, which resulted in a sharpening of animal senses and instincts and a degeneration of reasoning capacities, indicate that civilized Man, too, could descend into the ontological interzone of the "human creature" between humanity proper and human animality? As the article somewhat ominously concludes, "at what point, – or indeed, if at any point, – this [retrogressive] process at length stops short" is not easy to determine (p. 6317).

Between savagery and civilization

In its uncertain relation to savage and civilized humanity, the frontiersman complicated the convenient relegation of savage indigenes to the status of a "connecting link between the animal and intellectual creation," as explorer Jedediah Smith described the Indigenous people he encountered in the Sacramento Valley during his 1826–27 expedition to California (1977, p. 185). Smith's reference to a "connecting link" evokes the notion of the Great Chain of Being, which was still influential throughout the nineteenth century, both before and after (read: in opposition to) the rise of Darwinian evolutionary theory. The Great Chain was a supposedly fixed cosmological arrangement of hierarchical links reaching from inanimate matter and the lowest forms of life to humanity (with Man occupying its highest echelons) and the supernatural realms of celestial beings and God. The eighteenth and nineteenth centuries saw a continuous struggle to (re)draw the boundaries between humanity and animality in accordance with the divine "natural" order represented by the Great Chain. In the context of these endeavors to pinpoint human/animal difference, zooanthropological imaginings relied to a significant extent on precisely those "intermediate" forms that connected both realms while not clearly belonging to either. Occupying "a liminal position in a[n] . . . ontological scheme demarcating human from non-human" and thus posing "a distinct problem for the ontology and epistemology of separate being," the figure of the savage was crucial to these debates (Anderson, 2007, p. 40).

However, the discursive nexus of savagery and civilization was not always limited to a preoccupation with forms or stages of humanity but extended beyond the human sphere. Transgressing the conceptual pairings of savagery/animality and civilization/humanity, both terms were also employed to describe trans- or interspecific ontologies – ways of being that were not exclusive to humans or any other particular species but characterized by the interrelations between different kinds of creatures. In Crèvecoeur's writing, civilization is not exclusively embodied by Man, and Man not its sole (though certainly its most emblematic) representative, since the domesticated creatures under the care of the white farmer are an integral part of Crèvecoeur's civilized ontology in contradistinction to his depiction of frontier savagery, with the latter similarly involving the – in this case antagonistic rather than paternalistic – relations between a range of (mostly predatory) creatures, such as wolves, bears, foxes, and human hunters who have abandoned the civilized pursuits of agriculture. What civilized and savage life thus had in common was that both were composed of interspecies assemblages rather than merely representing the respective endpoints of a self-contained continuum of humanity. In the eyes of white Americans, the bodies and behaviors of animals and their relations with Indigenous humans functioned as indicators of civilizational status. This was particularly true with regard to animal domestication, which was seen as involving a process of human self-domestication that in fact constituted a crucial prerequisite for the unfolding of civilization itself. "Indian" dogs in particular became frequent objects of contempt and were unfavorably compared to their civilized counterparts, who accompanied their human "masters" westward. In *Astoria*, Irving is repulsed by the canines who "swarm about an Indian village as they do about a Turkish town" and have retained the "savage but cowardly temper" of the wolf, "howling rather than barking; showing their teeth and snarling on the slightest provocation, but sneaking away on the least attack" (1836, p. 223). Similarly, Parkman, encountering a group of Pawnee on the way to their hunting grounds, remarks on their "innumerable multitude of unruly wolfish dogs, who have not acquired the civilized accomplishment of barking, but howl like their wild cousins of the prairie" (1996, p. 60).

While antebellum discourse was often focused on the irreconcilability of savage and civilized ways of life, some contemporaries offered a more generous perspective on both the permeability of the boundaries separating savagery and civilization and on the notion of savagery itself. Catlin's "relativist" critique of the common usage of the term as something "expressive of the most ferocious, cruel, and murderous character that can be described," equates contemporary ideas about "the poor red man" with those about "feared and dreaded" predatory animals such as grizzly bears in order to make a point about the misuse of the concept of savagery as such. Both grizzlies and "Indians" may be wild, which, Catlin explains, is in fact the original meaning of the term *savage*, but this wildness did not inevitably translate to a bloodthirsty cruelty directed against civilized creatures and lifeways (1841, p. 9). And yet, while apparently all creatures could descend from the heights of civilization into a state of savagery, the same was not necessarily true the other way around. Perhaps Catlin was wrong and both "Indians" and bears were indeed fated to be enemies of civilization, an idea that also served to explain their "disappearance" in the face of civilization's westward march across the continent (cf. Brantlinger, 2003). "Boddlebak, the Bear-Tamer," a curious short story published in the August 1850 issue of *The Knickerbocker*, functioned as a cautionary tale for those who ignored, or willfully transgressed, the boundaries that separated both savagery from civilization and animality from humanity. The story tells of a young frontier settler who, while in pursuit of a bear in the Catskill Mountains, unwittingly wanders into the den of the eponymous antagonist, "a fierce-looking old man, with coarse white hair and grisly countenance" ("Boddlebak", 1850, p. 153). Looking around the dimly lit dwelling, to his surprise the young man finds, "seated in rough chairs made of oaken boughs, at least half a dozen bears," with the old man appearing to speak to the animals "in a discordant voice, using the same language which he might have addressed to intelligent beings" (p. 154). After he has chained the unsuspecting settler to a wall, we learn of Boddlebak's persistent endeavors to civilize "his" ursines, creatures, he admits, who may have "stood low . . . in the scale of beings" but whose inherent potentials and "great powers of mind . . . only required development." In fact, what the deranged old man seeks to prove with his experiment is the both absurd and dangerous belief "'that all are by nature the same'" (p. 155), that there is no boundary – metaphysical, linguistic, biological, or otherwise – that forever separates savage and civilized, animal and human. Unsurprisingly, while the bears at first seem to be "making progress in civilization" and the young man, imprisoned in Boddlebak's liminal domain ("half house, half cavern"), feels himself "becoming more and more like a beast," the cathartic ending of the story has the delusions of this "monstrous hater of his species" (pp. 155–156) shattered and his body torn to pieces by the apparently not-so-civilized bears. Set well before the era of expansionism during a period when New York's Catskills were still part of the "western" frontier, the story nonetheless communicates and negotiates antebellum ideas about the precarious boundaries and the struggle between savagery and civilization that played out in trans-Mississippi environments.

In the nineteenth-century West, the in-between of savagery and civilization was often addressed with the (usually but not exclusively) pejorative term "half-breed." While the term could also describe the characteristics of particular animals, such as "half-breed" horses combining the traits of the small, tough "Indian ponies" with those of American horses, it was mostly used as a reference to "mixed-race" trappers, hunters, scouts, and other human inhabitants of Western environments who were of both Indigenous and European descent, exemplars of the "rabble rout of nondescript beings that keep about the frontiers, between civilized and savage life," as Irving puts it (1835, p. 19). Irving's travel companion Charles Latrobe, elaborating on a visit to a Cherokee settlement near the Arkansas River where the party took up quarters with a "half-breed" called "Frenchman Jack," points to the way in which this concept, like the master concepts of savagery and civilization themselves, in fact frequently ignored the boundary between humans and animals to describe ways of being that encompassed and blended the worlds of both. A stark contrast to the orderly domain of the Crèvecoeurian farmer, the scene of "negligent thriftiness" that presents itself to Latrobe's eyes is animated by an ensemble of creatures whose liminal character seems to mirror that of their human companion(s) (or the other way around):

> fierce-looking pigs, with bristling mane, and erect, pointed ears, [who] eyed you a moment with straddling legs . . . , and then, grunting savage defiance, scampered away over the dead leaves. As usual, the dogs were numerous, and seemingly a cross between the dog and the wolf. Every thing had an air of half-breed, and from this the fowls were not an exception; the bodies of the hens were raised up upon long, yellow, unsightly legs to an unusual height, and a peculiar breed of ducks was not wanting to complete the picture.
>
> (1835, pp. 255–256)

Although whites sometimes commented favorably on the adaptation of both human and animal "half-breeds" to the specific demands and impositions of Western life and environments – the "half-breed" scout in particular is a ubiquitous figure in the myth-history of the West – human "half-breeds" were frequently depicted as combining only the worst qualities of savagery and civilization, testifying to the growing anxieties about miscegenation and racial degeneration associated with westward expansion and the encounters between different groups of humans it brought about (cf. Horsman, 1981).

Similar to the precariously civilized white frontiersman and the savage "Indian," inhabiting the ontological twilight zone between the two, the "half-breed" underlines how, on the one hand, contemporary zooanthropological and racial discourses, in their attempts at delineating the boundaries between savagery and civilization, animality and humanity, ironically seemed to rely on precisely such figures whose appearance, behaviors, and modes of existence threatened to further unsettle these boundaries. On the other hand, however, the situatedness of these figures at the geographical periphery of the nation and the ontological limits of Man meant that they were particularly well-suited to provide insights into such core issues as the fixity or malleability of human nature, the uniqueness of human moral and reasoning capacities, or even the single or separate origin(s) of human "races", because true insights into these issues seemingly demanded a conceptual and, sometimes, a physical venture beyond the frontiers of what contemporaries regarded as civilized existence.

Notes

1 For Indigenous histories of the trans-Mississippi West prior to the era of American expansionism, see, for example, Calloway (2003) and John (1996).
2 I use the term "zooanthropological imaginary" to describe a nexus of different and at times conflicting discourses and imaginings centered on the question of human–animal difference that was not limited to specialist debates in scientific circles but extended into the broader cultural sphere through the writings of (amateur) historians, novelists, journalists, travelers, and others.
3 For reasons of textual aesthetics and readability, I will refrain from putting these frequently used terms in quotes for the remainder of this chapter.
4 Biographical information about Webber can be found in *Handbook of Texas Online*, Thomas W. Cutrer, "Webber, Charles Wilkins," www.tshaonline.org/handbook/online/articles/fwe08.

References

Anderson, K. (2007). *Race and the crisis of humanism*. London: Routledge.
Anderson, W. M. (1967). *The Rocky Mountain Journals of William Marshall Anderson: The West in 1834* (D. L. Morgan and E. T. Harris, Eds.). San Marino: Huntington Library.
Atkinson, R.J.C. (1859). Reason and instinct. *The Zoologist: A Popular Miscellany of Natural History, 17*, 6313–6317.
Boddlebak, the bear-tamer: A legend of the Kaatskills. (1850). *The Knickerbocker, 36*, 151–159.
Brantlinger, P. (2003). *Dark vanishings: Discourse on the extinction of primitive races, 1800–1930*. Ithaca: Cornell University Press.
Calloway, C. G. (2003). *One Vast Winter Count: The Native American West before Lewis and Clark*. Lincoln: University of Nebraska Press.
Catlin, G. (1841). *Letters and notes on the manners, customs, and conditions of the North American Indians* (Vol. 1). New York: Wiley and Putnam.
Crèvecoeur, J.H.S.J. de. (1904). *Letters from an American farmer*. New York: Fox, Duffield & Co.
Cronon, W. (1995). The trouble with wilderness; or, getting back to the wrong nature. In W. Cronon (Ed.), *Uncommon ground: Toward reinventing nature* (pp. 69–90). New York: Norton.
Curti, M. (1980). *Human nature in American thought: A history*. Madison: University of Wisconsin Press.
Flores, D. (1999). The Great Plains 'wilderness' as a human-shaped environment. *Great Plains Research, 9*(2), 343–355.
Grossman, J. R. (Ed.). (1994). *The frontier in American culture*. Berkeley: University of California Press.
Hall, J. (1828). *Letters from the West*. London: Henry Colburn.
Hall, J. (1833). On Western character. *The Western Monthly Magazine and Literary Journal, 1*(1).
Hamilton, A. (1907). *Hamilton's itinerarium: Being a narrative of a journey from Annapolis, Maryland, through Delaware, Pennsylvania, New York, New Jersey, Connecticut, Rhode Island, Massachusetts and New Hampshire from May to September, 1744*. (A. B. Hart, Ed.). Saint Louis: William K. Bixby.
Horsman, R. (1981). *Race and manifest destiny: The origins of American racial Anglo-Saxonism*. Cambridge: Harvard University Press.

Hyde, A. F. (1990). *An American vision: Far Western landscape and national culture, 1820–1920*. New York: New York University Press.

Irving, W. (1835). *A tour on the prairies*. London: John Murray.

Irving, W. (1836). *Astoria: Or, enterprise beyond the Rocky Mountains*. Philadelphia: Carey, Lea and Blanchard.

John, E.A.H. (1996). *Storms Brewed in Other Men's Worlds: The Confrontation of Indians, Spanish, and French in the Southwest, 1540-1795* (2nd ed.). Norman: University of Oklahoma Press.

Latrobe, C. J. (1835). *The rambler in North America, 1832–1833*. London: Seeley and Burnside.

Lemay, J.A.L. (1978). The frontiersman from lout to hero: Notes on the significance of the comparative method and the stage theory in early American literature and culture. *Proceedings of the American Antiquarian Society, 88*, 187–223.

LeMenager, S. (2004). *Manifest and other destinies: Territorial fictions of the nineteenth-century United States*. Lincoln: University of Nebraska Press.

Lestel, D. (2014). Hybrid communities. *Angelaki, 19*(3), 61–73.

Limerick, P. N. (1987). *The legacy of conquest: The unbroken past of the American West*. New York: Norton.

Nash, L. (2006). *Inescapable ecologies: A history of environment, disease, and knowledge*. Berkeley: University of California Press.

Nash, R. F. (2014). *Wilderness and the American mind* (5th ed.). New Haven: Yale University Press.

Nugent, W. (1992). Where is the American West? Report on a survey. *Montana: The Magazine of Western History, 42*(3), 2–23.

Ohrem, D. (2017). The ends of man: The zooanthropological imaginary and the animal geographies of westward expansion in Antebellum America. In D. Ohrem (Ed.), *American beasts: Perspectives on animals and animality in U.S. culture, 1776–1920* (pp. 245–278). Berlin: Neofelis.

Olmsted, F. L. (1857). *A journey through Texas, or, a saddle-trip on the south-western frontier*. New York: Dix, Edwards & Co.

Outka, P. (2008). *Race and nature from transcendentalism to the Harlem Renaissance*. New York: Palgrave Macmillan.

Parkman, F. (1996). *The Oregon Trail*. (B. Rosenthal, Ed.). Oxford: Oxford University Press.

Ruxton, G. F. (1847). *Adventures in Mexico and the Rocky Mountains*. London: John Murray.

Scott, D. (2000). The re-enchantment of humanism: An interview with Sylvia Wynter. *Small Axe: A Caribbean Journal of Criticism, 4*(2), 118–207.

Slotkin, R. (1985). *The fatal environment: The myth of the frontier in the age of industrialization, 1800–1890*. New York: Atheneum.

Smith, H. N. (1970). *Virgin land: The American West as symbol and myth*. Cambridge: Harvard University Press.

Smith, J. S. (1977). *The southwest expedition of Jedediah S. Smith: His personal account of the journey to California, 1826–1827*. (G. R. Brooks, Ed.). Glendale: A. H. Clark & Co.

Turner, F. J. (1961). The significance of the frontier in American history. In R. A. Billington (Ed.), *Frontier and section: Selected essays of Frederick Jackson Turner* (pp. 37–62). Englewood Cliffs: Prentice-Hall.

Valencius, C. B. (2004). *The health of the country: How American settlers understood themselves and their land*. New York: Basic Books.

Webber, C. W. (1851). *The hunter-naturalist: Romance of sporting; or, wild scenes and wild hunters*. Philadelphia: J. W. Bradley.

White, H. (1972). The forms of wildness: Archaeology of an idea. In E. J. Dudley & M. E. Novak (Eds.), *The wild man within: An image in Western thought from the Renaissance to Romanticism* (pp. 3–38). Pittsburgh: University of Pittsburgh Press.

White, R. (1991). *'It's your misfortune and none of my own': A new history of the American West*. Norman: University of Oklahoma Press.

Wrobel, D. M. (1993). *The end of American exceptionalism: Frontier anxiety from the Old West to the New Deal*. Lawrence: University Press of Kansas.

Wynter, S. (2003). Unsettling the coloniality of being/power/truth/freedom: Towards the human, after man, its overrepresentation: An argument. *CR: The New Centennial Review*, *3*(3), 257–337.

11 For the love of life

Coal mining and pit bull fighting in early 19th-century Britain

Heidi J. Nast

There is indeed no class of persons, sailors themselves not excepted, who have greater reason to live in constant readiness to encounter death than the colliers who work in some of our deep and impure mines . . . under circumstances of excruciating trial.

(in *Quarterly Review* [QR], 1842, p. 187)

Introduction

There has been no political economic study of pitted dog fighting to date. For the most part, those writing about the blood sport address its history idiographically, drawing on interviews, fieldwork, 19th- and 20th-century dog-fighting anecdotes (primarily from the UK and US), and/or 19th- and 20th-century photographs, engravings, and paintings of dogs or dogs with their owners. This emphasis on describing as a form of truth-telling has permitted little in the way of theorizing *why* pitted dog fighting happened *where* and *when* it did/does, or why it has appealed to the *demography* it has and does. Instead, such description has lent itself to other kinds of truth-telling relayed in moralizing terms. Crudely put: pitted dog fighting involves a great deal of bloodshed that tells us that those involved are cruel, criminal, and bad; by contrast, those engaged in pitted dog rescue and reha-bilitation are kind, heroic, and good (e.g., Gorant, 2017, 2011; Dickey, 2017; Cum-mins, 2013; Foster 2012). Nineteenth-century language about the inherently immoral and cruel nature of dog fighting was first formulated by bourgeois reform-ers and continues to be recycled, except that amongst even the earliest writers there were those who recognized and took umbrage over the class-striated hypocrisy concerning how vice and crime were defined and punished (e.g., Proceedings, 1809). That said, a number of scholars have related contemporary pitted dog fight-ing to masculinity (a sport almost entirely engaged in by men), their work begin-ning to trace out some of the sport's historical material contours (e.g., Evans, Gauthier, Forsyth, 1998; Cummins, 2013; Smith, 2011).

This chapter picks up on these and similar impulses to explore the earliest political economy and geography of pitted dog fighting, based in Britain. Drawing heavily on primary and secondary sources, I explore how the Industrial Revolution, and early British colliery life, specifically, produced material equivalencies between pitted dog fighting and pitted men, both of whom were made to operate "under circumstances of excruciating trial." The bull and terrier dogs that became the lifeblood of pitted dog breeding were colliers' most intimate companions, these dogs' documented abilities to fight unprovoked modeling how colliers would come to see themselves – namely, as fighters uniquely able to survive conditions not of their own making.

Of special importance are the findings of the Parliamentary Children's Employment Commission that Queen Victoria established after the famous 1838 flooding accident in Yorkshire's Huskar pit that killed 26 boys and girls between ages 7 and 17. Dozens of experts were sub-commissioned and dispatched to study colliery life across Great Britain, each one writing a separate report, many of these filled with ethnographic details and accompanied by sketches. The reports were compiled and published as a lengthy Appendix to the *First Report of the Commissioners (Mines)* in 1842. The publication of the more than 1,000-page Report and Appendix was historic, providing fodder not only for social commentators (including Marx and Engels) and writers (Dickens and Mayhew) but also for reformers and reformist legislation. The first legislation to follow was the Mines and Collieries Act of 1842, which outlawed children under the age of 10 and all women from mine work, persons whose wages had always been a tiny fraction of those given to men.

The sketches and descriptions are important as they demonstrate how the earliest colliers labored in positions and under conditions most commonly associated with work animals. Child laborers who worked in especially thin seams, for instance, were regularly placed in dog chains and harnesses, while workers were commonly carried into and out of the mine shafts via large metal chains and, eventually, metal buckets and cages. The powerful visual and anecdotal evidence that the Report and Appendix provides, then, helps to establish the profound spatial and bodily resonances between dog pitting and human pitting.

The chapter also examines how strongly male colliers identified, and were identified by others, with fighting dogs. Their affective identification allows me to trace how early dog fighting – like bare-knuckle boxing – incarnated and carried forward the competitive and masculinist logic of industrial capitalism. The crudeness of early technologies, the relative isolation of the mines, and mine owners' desires to maximize coal production and profits gave way to unprecedentedly high mortality rates. Dog pitting accordingly dramatized, perhaps most clearly, the competitive forces that similarly pitted and mangled the human. Yet, pitmen did not see dog fighting as dramatizing their own social or economic castration but as a scene for *recognizing* and *validating* the hard work that they did. As such, the pit operated unconsciously, providing a compensatory space for psychically organizing and managing early colliery life. For pitmen, fighting was about improving the odds of life, the musculature of man and dog made to speak of victorious possibilities rather than defeat (c.f. Davies, 2011).

The pitted bull and terrier

The first pitted bull dogs – or pit bull dogs – came from the mixed mastiff-alaunt stock that British farmers had used as working animals and in bull-baiting events, beginning in the late Middle Ages. The best bull-baiting dogs had a low center of gravity, a good prey drive, and slightly flattened snouts, the latter allowing them to breathe while biting and holding onto the bull. Otherwise, bull dogs varied in appearance.

Rural bull-baiting, like the pit(ted) bull dog fighting that would follow, embodied the political economy of the time: the landed nobility (or the church) owned and provided the bull while farmers supplied the dogs; the species-inflected differences in animal size and value spoke to the class distinctions of the owners. That said, both animal kinds were equally well regarded, in large part because the outcome of bull-baiting was far from certain. The bulls were admired for being resourceful, strong, and smart, whereas the dogs were known for their tenacity, agility, and courage.

The popularity of bull-baiting waned for many reasons: (a) the late 18th- and early 19th-century Parliamentary enclosures that broke up feudal institutions and eliminated the commons; (b) the encroachment of factory towns and cities; (c) the related disappearance of the large public rings and open spaces used in baiting; and (d) the 1835 Parliamentary Cruelty to Animals Act, which made all forms of animal baiting and fighting – including that between dogs – illegal (Griffin, 2005). Whereas the first several pressures were tied to growing agro-industrial, industrial, and commercial interests, the Act was borne by new animal welfare sensibilities contradicted by (among other things) the rise of big game hunting, the eugenics of standard-making in the breeding of pets and livestock, and the widespread enclosure of animals into domestic spaces, public zoos, and scientific laboratories (Rouse, 2015, p. 89; Griffin, 2005; Ritvo, 1986; Watts, 2000). The Act therefore punctuated (rather than caused) bull-baiting's end, its understanding of cruelty irrelevant to the necropolitics and commodification coming to govern human and nonhuman animal lives. While the large size of bulls made it difficult to evade the 1835 Act, the comparatively small size of dogs made evading its proscriptions easier. More important, however, is that dog fighting, unlike bull-baiting (or ancient and contemporary cockfighting), involved the pitting of *competition between equals*, using animals with which "dogmen" *expressly identified and considered to be their most intimate and constant companions*. As such, dog fighting lent itself to, and *gained* vitality and meaning from, the competitive profit-seeking vicissitudes of industrial capitalism. Such vitality was nowhere more evident than in the rural coal mining areas, where I argue the sport first effloresced, and in the British cities where it later traveled and expanded. If anything, the Act *re-directed* the sport underground and provided the legal basis for racialized and class-inflected moralizing.

That dog fighting traveled to cities early on is evident in the writings of the English ex-convict, William Derricourt (b. 1819), which the Australian writer, Louis Becke (1855–1913), serialized and published in 1899 as *Old Convict Days*. Derricourt begins by recounting his apprenticeship with a local gunlock filer and publican in Darlaston, a small impoverished town located on the South Staffordshire coalfield and known for its nut, bolt and gunlock manufactories. Despite not living in a colliery, his employer was well acquainted with dog fighting, the intricacies of which he taught to William early on:

> [b]rought up in this school [i.e., being taught how to train dogs], and having it always dinned into me that *the man who keeps a fighting dog must be a*

fighting man, it is very little wonder that I got some of the nature of those around me – *men and dogs alike* [emphasis added].

(Becke, 1899, p. 3)

Elsewhere, Derricourt makes plain the extent to which working-class men identified their lives and struggles with those of fighting dogs. This equivalency is evident in his description of "French and English," a working-class bettor's game where two male youth are hoisted onto older working men's shoulders and made to fight a match "to a finish" (Becke, 1899, p. 3). Having scrapped for much of his life, William proved fight-worthy and became "a great favourite among the fighting people" (in Becke, 1899, p. 4). It was additionally seen in rat-killing competitions, where working-class men (kneeling, with hands tied behind their backs) were placed successively in pits to compete with one another using only their teeth, dogs often set to the same task; and it was there in the twice-yearly bare knuckle-fighting matches staged between roughly 30 youth each from Darlaston and the nearby town of Wilnon. As in dog fighting, the latter event saw "the loss of much blood and skin . . . [and] sometimes . . . fatal injuries. Boys were bruised and stunned. . . . After the scrimmage the wounded were wheeled to their homes on barrows" (in Becke, 1899, pp. 4–5, the same way dogmen in the 19th-century US took unsuccessful fighting dogs home (see Nast, 2015).

The fights of men and dogs thereby came to be similarly staged and identified with one another. Questioning why this was so requires addressing how male colliers saw pit work as a kind of battle for life.

Coal pits, dog belts, and "chains"

Sometime in the 1840s, Britain's largest coal and iron ore mining district (extending through what was then southern Staffordshire and northern Worcestershire counties) became known as the "Black Country."[1] Limited surface mining for individual or collective use had taken place there since the 14th century; however, after surface outcrops were depleted, miners began digging shafts to flare out toward the bottom within the earth, creating the so-called bell pit, which was unstable and prone to collapse (see Youles et al., 2008). While coal demand increased by the mid-16th century as forestland was depleted, its utility was still limited. This changed in the latter half of the 18th century as coal became the main energy source for the Industrial Revolution and new technologies made deep pit mining possible. Coal mining subsequently intensified and expanded across the West Midlands and other coal mining areas of Great Britain, including Wales and Scotland, where it had already long been practiced. Its intensification led to the massive in-migration of workers and a mass out-migration of the gentry. As one of the Report's sub-commissioners assigned to the Black Country averred, the gentry reviled these migrants not only because they were harbingers of an industry now destroying the "face of the country," but because of their visible poverty, reduced to an aesthetic:

> numerous ugly cottages spring up like a crop of mushrooms – long rows of wagons, laden with ill-assorted furniture, are seen approaching, and with them the pitmen and their families. This is the signal for the departure of the gentry, unless they are content to remain amidst 'the offscouring [rubbish] of a peculiar, a mischievous, and unlettered race' . . . to see their district assume a funereal colour – 'black with dense volumes of rolling smoke,' and echoing with the clatter of endless strings of coal-waggons [sic].
>
> (*QR*, 1842, pp. 160–161)

Prior to the invention of surer technological means of building, stabilizing, illuminating, draining, and ventilating deep pits, deep pit coal mining in the late 18th and early 19th centuries was the most dangerous of industrial professions. Working inside them meant chronic exposure to fine airborne particulates, and, in the deep seams of Britain's northeast where gas pressures were especially high, explosions were common. Public outcry over the frequency of mining accidents and the findings of the 1842 Report eventually led Parliament to enact the 1850 Coal Mines Inspection Act. Four mining experts were appointed Inspectors of Coal Mines and tasked with collecting and publishing mine-related mortality data and inspecting and elaborating coal mine safety standards, though the number of Inspectors soon swelled. Historian P.E.H. Hair (1968, p. 548) analyzed these Inspector reports alongside earlier data to derive what collier mortality rates might have been between 1800–1850, when there was no systematic record-keeping. He concluded that mortality rates amongst collier men were four to ten times higher than for the general male population, depending on region and year (he mentions nothing about women). This accelerated death rate meant that by the 1840s, "only half as many old men above [aged] seventy [existed] among colliers as among agriculturalists" (*QR*, 1842, p. 191). If the mining-related deaths of children under the age of 10 had been taken into account (these were never recorded), these figures would be even worse.

Early on, children as young as seven or eight years old were deployed as trappers, whose jobs it was to operate a pit's ventilation system. They started work daily at 2 a.m. at the above-ground mine entrance or pit-head, coffee tin and bread in hand, two hours earlier than the adults. From here, they were sent to specially built dug-outs adjacent to shaft passageway doors, which in small seams could be very small. Their main job was to pull the door open with a string upon hearing a burrower approach, allowing him to pass and deposit his load in a collection site nearby. The opening and closing of these doors also operated as a kind of primitive valve system for regulating pit pressures and temperatures. Many trappers worked sitting on their haunches in total darkness for up to 12 hours a day.

Children also carried out transport-related tasks that capitalized on their small size.[2] The youngest and smallest of these were putters. Putters pushed or pulled

Figure 11.1a A young female putter in a dog harness and chain pulls a coal corve

Source: From the British Parliamentary Papers on Children's Employment, Royal Commission Report, Reports and Evidence from Commissioners, volume 7, Appendix 1, 1842

shallow wooden transport boxes (corves) with or without runners filled with hundreds of pounds of coal through shafts hundreds of feet long (Figure 11.1a). As a result of the excessive heaviness of their loads, putters' arms and/or legs were often bowed. In Shropshire and other counties, where coal seams were less than 24 inches thick with little expanse of rock in between, the children worked literally like dogs: their waists were fitted with leather dog-belts and harnesses attached to sturdy chains, which allowed them to pull the corves while crawling on all fours. Because owners were most interested in maximizing coal extraction, no protective lining was placed along these seams, with the result that many putters' knees and backs were scarred. While some larger mine owners had replaced this extraction method with mechanical means by the time of the Report, many smaller mine owners had not (Chawner, 1842, p. 42).

Children were additionally treated like work animals through the "butty" labor contracting system. Butties were a special group of working-class men who received a share of mining profits in lieu of wages, this incentivizing them to become part of the exploitation process. To secure the cheapest and healthiest labor, butties raided the workhouses of Lancashire, Yorkshire, western Scotland, and, especially, Staffordshire, for orphan and pauper children between the ages of eight and nine years old, not unlike how dog fighters picked out the healthiest puppies of a litter. While, technically, these children were to be released after "apprenticing" with a butty for 12 years, they effectively served as slave labor and suffered accordingly. With no one to advocate for them, they were given the most burdensome work, mistreated by other workers, harshly disciplined by supervisors, and deprived of adequate food. Their exhaustion was said to account for why at least some fell asleep on the pit floor and were unwittingly trampled by coal wagons (Chawner, 1842, pp. 60–62,

Figure 11.1b A flash photograph from 1930 showing one of the various kinds of metal chain-
ings used to raise and lower coal miners into and out of the coal pits. This is
the first coal pit photo taken using the Sashalight camera, its flash powered by
a battery, rather than an explosive powder. By this time, the chaining device
had become obsolete and was used only for emergency purposes

Source: Photo by Sasha/Getty Images

81–82). The most offensive butty practices took place in Walsall, Wolverhampton,
Dudley, and Stourbridge, part of the old Staffordshire region, which is coincidentally where pitted dog fighting took on special importance.

The working conditions for adult miners were not much better. Most started
their workdays at 4 a.m. by gathering at the pit-head. If the pits were relatively
shallow, as in the case of Staffordshire and (West Riding) Yorkshire, the workers
climbed onto skips or baskets that held four or more persons at a time and were
lowered via a thick rope into total darkness. In Shropshire and Staffordshire, the
workers were lowered by individual chains, instead. In this case, the metal yoke
and chain were disconnected from the skip and re-hooked to a series of looped
chains into which the miners stepped, "like a boy in a rope-swing" (Chawner,
1842, p. 6; Figure 11.1b). Up to 20 miners could ride up and down a shaft in this
way, provided that some of the younger boys and girls rode astride an older miner's
thighs. Upon reaching the bottom, candles were lit and:

> a new world is opened: – there are roads branching out for miles in every
> direction, some straight, broad, and even, others undulating and steep, others

narrow, propped by huge pillars' the whole illuminated, and exhibiting black, big-boned figures, half-naked, working amid the clatter of carriages, the incessant movements of horses, the rapid pace of *hurries*, the roar of furnaces, the groaning and plunging of steam engines.

(QR, 1842, p. 89)

Given that deep pits had inadequate drainage systems, flooding was commonplace and miners commonly worked in standing or brackish water. Rats and mice were commonplace, introduced through the hay feed brought in for the pit ponies, and the heat could be extreme. To cope with the latter, men and women regularly stripped down to the waist, and many men worked in their underwear. Sex within the pits was not unusual, nor was rape, opportunities for these having somewhat to do with how certain kinds of mine work required independent work teams of two (Chawner, 1842, pp. 77–80). Lastly, the ambient coal dust in combination with the dampness of, and long underground hours in, the pits caused most miners to contract pulmonary diseases while in their twenties and thirties. Workers in the West Midlands' counties of Shropshire and Staffordshire were thereby incapacitated by the age of 40, the same as in Derbyshire, though Warwickshire laborers could often work until age 50.

Despite the exploitative circumstances, male miners were known for their drinking (Staffordshire mine owners paid workers partly in beer), combativeness, and labor organizing – especially in Wales. Non-mining townsfolk were said to fear for their person and property when these men came off their shifts or were off on Sundays, Saint Mondays (absentee days that miners often took), and holidays. One of the contributors to the 1842 Parliamentary Reports noted that: "[w]henever from any causes . . . the collier is '*unchained*,' [emphasis added] the police are on the alert for scenes of riot and fight" (*QR*, 1842, p. 172). The wording not only alludes to how chains were used to lower and raise colliers into and out of the pit, it suggests that workers unchained naturally gravitated to fight, an association with chaining repeated in dog fights. Police records registered the truth of collier violence, which became "so certain, that the police have only to know that colliers are *unchained* from their ordinary work, either by *strikes* [emphasis added] or other causes, to be prepared for extraordinary outrages on persons or property" (Chawner, 1842, p. 77).

The idea that colliers were natural fighters had racial and sexual proportions. One of the Report's sub-commissioners writes of the "broad . . . stalwart . . . [and] swarthy collier" who doesn't simply go home after work, he "stalks home, all grime and muscle," a depiction both anxiety-ridden and admiring. The author contrasts the collier body with that of, "the puny, pallid, starveling, little weaver [textile worker], with his dirty white apron and feminine look" (*QR*, 1842, p. 189), reducing the bodily deformations involved in both cases (starvation and hyper-muscularity) to aesthetics, disappearing how either body was made (see also Metcalfe, 1982).

While discourses of the fighting dog and fighting worker came into alignment through the spaces of mine work (e.g., dog belts and chains) and the mining body (e.g., fighting titans enchained), such leveling was also seen in how the pitmen themselves identified with dogs. This had not only to do with the exigencies of pit work but from the fighting nature of the bull and terrier dogs, both of these becoming the business of pubs and pub owners.

Paydays, Sundays, and pubs: the cultural materialism of a pitted sport

As noted previously, the first pitted bull dogs – or pit bull dogs – came from re-titrating the mastiff-alaunt mixes that British farmers had used for bull-baiting. It was from the "sport" of bull-baiting that the name "bull dog" derived, it never having been a standardized breed. Pitting dogs against one another required something more than the tenacious grip, high prey drive, and special courage of bull dogs. Fights between equals required speed, agility, and a biting style that involved shaking and holding an opponent as well as a sense of when to change "to a fresh place of attack" ("Stonehenge," 1872, p. 162). All of these characteristics were secured by cross-breeding bull dogs with various terrier kinds. Primary data suggest that the category "Bull Terrier," or Bull and Terrier, was already in currency in dog-fighting circles by the early 1800s, though the uninitiated would continue to call them bull dogs for decades to come (Haynes, 1912, p. 712; Watson, 1906, p. 449). Scottish naturalist Captain Thomas Brown asserted as early as 1829 that the bull terrier had "assumed [such] a fixed character" that it deserved its own scientific name; accordingly, he assigned *Canis pugilis*, or, the fighting dog (p. 404).[3]

Male colliers cherished their bull terriers and many, if not all, households owned one. Like the farmers' bull dogs, these were working animals, protecting households, killing vermin, and often accompanying the miners wherever they went. While Sundays were favored for dog fighting, the blood sport could take place on any day after work. It also occupied collier men on their fortnightly payday – typically a Saturday – which mine owners treated as a holiday. Public houses became pivotal in this regard in that mine owners used them as a locus from which to disburse workers' wages, an activity carried out by his mine manager or butty, who typically also owned the pub! Whether the owner paid his workers in local currency, gold, or bank notes (the latter two needed to be exchanged for silver to be legal tender), the mine manager combined his positions in ways that not only directly benefited him but also made the pub into an institution vital to the growth of pit culture.

The pub's importance had much to do with how the butty combined his role as publican to his advantage. On paydays, for instance, he regularly announced a time for the owner's arrival that was much earlier than was the case. In this way, he not only encouraged the colliers to drink, but he did so knowing that upon the owner arriving, he could directly dock their pay. Moreover, as a businessman with some liquidity, he had the means with which to exchange bank notes or gold with silver, this allowing him another opportunity to exact a fee (Chawner, 1842, p. 85; *QR*, 1842, p. 168).

In orchestrating this extended period of drinking, the pub became a space of entertainment, one part of which involved gambling and blood sport. Pub owners encouraged these activities not only by allowing them to take place in and near the premises, but also by regularly sponsoring fights and providing prize monies (c.f. Metcalfe, 1982). This anchoring would become especially important after the 1835 Cruelty to Animals Act, when public houses across Great Britain helped to sustain the sport.

The dogs that colliers used were the same ones that guarded their households and were used in the informal and formal matches that regularly took place on the streets and open spaces of colliery towns and villages. A surgeon working in a Lancashire colliery reported that in the summer, colliers would sometimes, "sit all round the door of the public-house in a great circle, all on their hams, every man his bull-[terrier] dog between his knees; and in this position they will drink and smoke" – exactly how young trappers sat and how "dog men" in some places were expected to hold their dogs immediately before releasing them to fight (*QR*, 1842, p. 169). Another pub-related account given 60 or so years later relays how dogs taken inside the pubs might spontaneously start fighting if located too close to one another while at their owner's feet:

> The moment a dog discovered he was within reach of another dog, he dived right in. . . . Any vacant stools within yards spun off, apparently of their own volition; glasses were nudged from fingers and crashed from rocking tables. In their effort to steer clear of the canine whirlwind among their feet, several men toppled to the floor, their dogs escaping in the confusion, to join the fray.
> (Drabble in Cummins, 2013, p. 167)

Dog fighting also took place near or in the pits themselves. According to an amateur historian and resident of the Black Country, "it was not uncommon for men to go to work with there [sic] pit bulls or there fighting cocks for a spot of sport on their break times," a practice that likewise obtained in the Welsh-dominated coal mines of Pennsylvania.[4] Through these spatial relationships clear metonymic associations came to obtain between pitmen, dog fighting, and pit mining, associations cemented by the many pit-related terms that defined the spaces of colliery life, such as pit head, pit shaft, pit brow, pit bank, and so on (Gresley, 1883).

Many outside coal mining areas used collier fascination with blood sports, particularly dog fighting, to disparage them as prone to vanity and vice. Henry Morton, Esq. (a notoriously inhumane manager of the Earl and Countess of Durham's estates and collieries) commented in 1841 that: "They [the pitmen] consider themselves vastly superior . . . to agricultural labourers. Drunkenness is a prevalent vice, and dog-fighting is a favourite amusement" (in Great Britain, 1842b, No. 386). That dog fighting was done mostly on Sundays bolstered claims of collier immorality, even though it was the only official day they had off, at least until the 1870s. Such moralizing twined easily with the records of the police. Hence, the author of an 1846 Parliamentary Report on schools in Britain's coal mining "Northern District" notes that:

> [in] the coal-field of Durham and Northumberland, the Sabbath-day . . . is a strange scene in a Christian land. There is often no place of worship. It is a

day of profane and noisy pleasure; the men are lounging about; some in their usual, not working-dress. . . . In one part there is a dogfight. . . . On all sides you hear the ring of quoits, and the brutal language and coarse jeers of those who are playing with them.

(Watkins, 1846, p. 161)

This "unChristian-ness" was used to malign the miners politically, especially in relation to labor strikes. For example, on 1 February 1873, the *London Illustrated News* ridiculed a Welsh collier strike with an image titled, "The Strike In South Wales: A Sunday's Amusement." In it, ill-suited miners and their followers ballyhoo two pitted dogs in their circling midst. Sunday – the title ironically announces – is when good people go to church. This framing of the winter strike as entertainment – and in a form moralistically eschewed by the paper's readers – allowed the newspaper to cast Welsh miners' concerted efforts as a form of brutish play, one afforded by sacrificing everyone else's warmth. Two weeks later, upon hearing that colliers objected to the image, *Punch* (15 February, p. 66) insulted the strikers further, reporting that what *really* worried them was that readers might think they fought dogs *only* on Sundays.

Sunday's importance, however, begs a somewhat obvious question: why would pitmen, who had just finished a grueling workweek, use their official day off not for rest and repair, but for staging violences matching in intensity and architecture those they had momentarily escaped. Was the compulsively repetitive staging of blood matches between equals a means of containing, enunciating, and displacing the traumatic contours of collier lives? A way of dramatizing the competitive logic shaping the world around them? If so, the theaters they created and the blood-related rules they followed were remarkably in keeping with the competitive tenets of industrial capitalism, ones that in the 1860s would be re-articulated in social Darwinist terms.

The tight relationship between colliers, dogs, and dog fighting, while forged locally in individual (and quite often, insular) mining communities, assumed larger social geographical significance when colliers started organizing regional events with nearby mining towns. Unlike in the past when the winning dog stood for his owner, here winning assumed larger proportions. The winner stood for both his owner and his owner's colliery, the competitive blood sport easily enfolding into itself the industrial ethos of coal mining (see Waugh, 1855, p. 190; Griffin, 2005; c.f. Metcalfe, 1982, p. 477).

Dog fighting gained even greater purchase by the turn of the 19th century, when the sport traveled into nearby cities where men of all social ranks began investing in it. This broadening had partly to do with the propinquity of many industrial cities and towns to collieries and the mid- to late 18th-century building of an extensive canal system that connected coal areas to factory and trading cities. But propinquity does not explain its popularity. This, I contend, had to do not only with men's affections for their dogs or their dogs' willingness to please. Rather, it had to do with the resonances between pitted competitions and the increasingly competitive world in which men found themselves. That is, dog fighting worked as a vernacular through which men of disparate classes could identify and place themselves. London elites, for instance, identified with it as a *game* on which to gamble – a pairing of risk assessment and investment that rehearsed what any venture capitalist would do. The privateers and slave merchants of Liverpool (the city accounted for 80% of all British slave trading in 1807), meanwhile, would have found vindication for their own cruelty- and profit-driven professions (Shimmin, 1856). The lives of the dogs, in either case, mattered relatively little. Most colliers and many other working-class men, by contrast, identified with the dogs; it not being uncommon for the death of a well-known "warrior" dog to become the basis of some collective mourning or for the birth of pups from a fighting pair to be a source of shared elation.

Dog fighting, in other words, allowed men, standing or sitting side-by-side, to imagine that they lived in a naturally competitive world common to them all. This, despite the fact that they held markedly different bodily and proprietary stakes in the games and the fact that it was through their inequalities that the pitted cruelties made sense.

The pit as world

The complicated ways that pit bull fighting traveled out of colliery life into all of the British Isles and, eventually into the US and British colonies, is an expansive topic that cannot be addressed within a single chapter. In short, the historic levels of migration and urbanization accompanying the Industrial Revolution, and the related cultural importance and pervasiveness of working-class pubs, facilitated the blood sport's geographical circulation and dispersion (see also Cummins, 2013). The bull terrier, itself, was also partially responsible; its abilities to catch and kill animals that were deemed pests were seen as vastly superior to those of terriers, while its related natural propensity to fight was well known. By the 1820s, aristocrats, industrialists, university students, factory workers, politicians, clergy, dock workers, pugilists, liverymen, "dandies," "criminal elements," the homeless, and the unemployed – almost entirely men (though prostitution was common) – were placing bets according to their means and for their own class-inflected purposes. This remarkable demographic breadth remained largely intact throughout the 19th century despite reformist interventions (see Egan, 1821; Shimmin, 1856; "Stonehenge," 1872, p. 131). Even bourgeois dog owners, especially in London, took to the bull and terrier, not for fighting purposes but for their fighting *look* or, as one sportsman opined, as a "fashionable appendage" (Brown, 1829, p. 405).[5]

As factory towns and cities became vital nodes for the transferring and broadening of dog-fighting culture, geographically based dog-fighting rules and regulations followed. Their formal elaboration may have started as early as 1800, each set named according to the place from which it derived. Amongst the most well-known were the Rules of Birmingham, London, and Yorkshire. Birmingham had long been a renowned center of scientific discovery and specialized metalworking, having only decades earlier been connected to the coal mining areas of the Black Country via a network of canals. London was the world's largest and wealthiest city, known for its commerce and industry as well as its large migrant population and tremendous poverty and wealth. And Yorkshire County was home to Sheffield and centers of coal mining, steelmaking, and textile production. The shared economic interests of, and throughways between, these places suggest that dog fighting's logic had found much broader industrial purchase.

That rules and regulations needed to be written down presumably had to do with a number of factors related to its expansion. First, betting purses would have been much larger in that they were no longer only the purview of local colliery pubs. Second, many participants were new to the game and would not have been equally well informed about how pitted dog fighting worked. Third, the rules stipulated pit dimensions, the fight moves allowed, and the point system in play, allowing even novices to set up a practice. Fourth, as investments intensified, new professional categories may have emerged, the Rules including rather lengthy lists of offices that included trainer, time keeper, stakeholder, referee, sponger, and "key official." While the trainer worked to prepare a dog physically for "battle," the key official licked the animals immediately prior to release to test if their skin had been tainted with irritants or poisons.

The London rules may have been written to coincide with the opening of London's Westminster Pit circa 1800, the most celebrated urban arena for dog fighting that was also used for ratting competitions and "Old School" large-animal baiting, in keeping with the city's Elizabethan past (Borrow, 1851; Figure 11.2). The word "pit" is instructive and refers partly to how these larger arenas were built as large rectangular enclosures or containers. The Westminster Pit was typical. Its fight floor was of regulation dimensions (18 feet by 20 feet) and surrounded by a thigh-high wall over which potentially hundreds of participants could peer, their numbers accentuating the pit's cavernous look. Such intimacy between onlooker and fight was strategic in that, unlike other forms of theater, gambling and prize money were involved, compelling greater investment scrutiny. By 1825, the Pit's owner had added a second tier of seating located straight-up above the pit floor wall, a verticality that added to a sense of the pit's depth (Figure 11.2).

The dogs that appeared in these pits would already have been "tried" at any number of informal and smaller formal contests. This allowed owners and potential investors to judge a dog's gameness and fight drive well ahead of time. If a dog whimpered (a sign of "the white feather"), it was culled (killed). Most fights were of these lesser vested kinds. The formal matches that took place in regulation pits, such as those in London, were different and involved a protocol. This began when both parties met to draw up and sign an official contract and tender a deposit with the official stakeholder. The dogs would then be weighed to make sure that they were in the same weight class, and a referee was appointed. Only then would the sides decide on a date and fight location. Once word spread of an impending fight, battle monies (bets) were largely collected "in driblets at the bars of *different* [emphasis added] public houses [pubs] week by week" ("Stonehenge," 1872, p. 129; see also Shimmin, 1856, p. 80).

Regardless of regional variances in fight rules, a formal "scratch" game required that the combatants be positioned in opposing diagonal corners – often blindfolded and held tightly between the thighs of its second. Upon the fight being called, each

The WESTMINSTER PIT.

Figure 11.2 One of the earliest images of London's Westminster pit, a hand-colored etching
and aquatint by Isaac Robert Cruikshank, 1820. Later images show that, by
1822, a second tier of seating had been added vertically by attaching shallow
benches and guard rails onto the side posts. The posh chandeliers provide light-
ing that in meaner circumstances simple candles would provide. This curious
use of candles and darkened rooms even before the 1835 Act – when such things
were precautionary – is highly suggestive of the coal pit.

Source: From the British Museum, online photo archive, open access

second removed his dog's blinds (of used) and released it into the pit, the dog now
being off the chain. Before the game could be considered a formal match, however,
each dog had to cross the "scratch" line, a line drawn in chalk at the pit's center –
or a string. A fight round was completed when one of the dogs "turned" its head
away, either to breathe or to withdraw. At this point, both dogs were taken back to
their corners for a time out, which might last only a minute, caretakers using this
time to sponge the dog off, treat its wounds, and encourage it to continue.

   The large purses made possible by the sport's growing popularity seem to have
changed the tenor of colliery dog breeding and fighting, at least by the 1870s. By
this time, miners' lives had changed for the better, mostly as a result of unionizing
in the 1830s. As of 1844, for instance, a colliery owner could no longer bond a
pitman to his operations for a full year, a kind of servitude that made pits into a
special kind of proprietary enclosure. And in 1872, Saturday became a regular

holiday, a momentous change that "permitted the real beginnings of modern organized sport in the coal field" (Metcalfe, 1982, p. 475). These changes would have made it more difficult for the "fighting" pitman to identify himself with and through the "fighting" dog, as he once had, especially given the large purses being gathered for pitted fights beyond the coalfields. This shift may help to explain why the well-known London surgeon and sportsman, John Henry Walsh (alias "Stonehenge"), felt it necessary to inform his readers mockingly about how much colliers invested in their fighting dogs:

> He [the fighting dog] has the best of meat – legs of mutton, even – milk, jellies, often enriched by a little port wine, cow-heel, and boiled bullock's nose, which is tough, and supposed to strengthen the jaws. Nothing is too good for him, and there is little exaggeration in the Black Country tale of the collier who asks his better half – "what have you done with th' milk?" "Gen it th' child!" "Why da'int yer gen it th' pup?" The "pup," with a stake of five or ten pounds impending, is of far more importance in the home than the child.
>
> ("Stonehenge," 1872, p. 129)

This statement, like others (for instance, Shimmin, 1856, p. 78), while perhaps exaggeratedly reformist in tone, is remarkable for at least two reasons. First, it points to what seems to have been a greatly improved collier diet! But, second, it is being delivered nearly 40 years *after* the 1835 Cruelty to Animals Act. Indeed, as many writers have recounted, the Act propelled pitted dog fighting into more isolated spaces, such as barns and cellars in rural areas, and smaller pits in working-class pubs and elite taverns. The lengthy comments of one disgruntled "canine tavern" customer are worth partly repeating here, in that they point to how working-class dogmen, in particular, were well aware of the class politics involved in dog fighting's criminalized demise:

> What with driving a fellow from fighting his dog in the fields or in a yard to fighting it in a [public] house, I don't know what things will come to; now a fellow can't fight his dog even in a house, in *ever so quiet a way* [emphasis in original] but an infernal "peeler" [cop] comes and shoves his smeller [nose] in. A swell [rich person] may ride his horse to death in the face of day, may break its neck or leg, and kill it on the field at a steeple-chase or race-course, whilst hundreds of other swells are looking on and enjoying the sport, yet nothing is said or done; but if a poor cove *gives his dog a turn* [emphasis added] for half an hour – it may be only for fun, or may be *to keep him in health* – a lobster [authority] will come and collar him, and away he goes to quod [jail], and the beak [judge] says, "Oh you brutal fellow, this is dreadful sport. You are a vile, disgraceful wretch, and must pay twenty shillings and costs."
>
> (Shimmin, 1856, p. 76, emphases added)

That the men held great affection for their dogs and that the dogs loved their owners is partly evident in the man's concern that a dog needs to have "a turn"; and that fighting, like any pugilist knows, is how the fighter (bull and terrier) keeps in shape. Affection is also there in Shimmin's description of two men playing cards at a pub table. One man's brindled bull and terrier sits contentedly on his owner's lap but constantly interrupts the game by licking his owner's face and hands, something that the owner takes in stride.

While after the 1835 Act many policemen looked the other way, many others did not. Those caught could be fined and, if a pub owner was involved, his pub could be shut down. Thus, in Liverpool, pub owners kept the whereabouts of a fight secret until the night before, rotating the locations of the games amongst themselves. They also charged an entrance fee purportedly to pay the publican's fines, should the place be raided. The timing of the games was changed to very early on Sunday mornings, before the police started their rounds, and a special knock was required before the pub door would be opened. Smaller pop-up pits were devised (8 feet to 10 feet square) that could easily be hidden or stored. Made of wood boards, it was "so arranged that it can be taken to pieces at a moment's notice and put aside, or more frequently into a closet which is at hand" (Shimmin, 1856, p. 80). As in the past, prizefighters were key organizers and players in these games, their shared identification with prize-fighting bull and terriers evident, in this case, by the portraits of both hanging side by side on one of the pub room's walls.

More than 15 years later, "Stonehenge" would similarly describe precautions that dogmen took to circumvent the law:

> Doors are barred, windows blocked up, and every aperture closed. No person can quit the place under any circumstances until the fight is over; the temperature is often tropical, and men strip to the shirt, and sit bathed in perspiration and half fainting for hours together. A few rats and a terrier are generally at hand as decoys, so that if a police raid should take place, the canine combatants would be stowed away somewhere, and the officers merely drop in upon a part of men mildly engaged in killing a few rats.
>
> (1872, p. 130)

Conclusions

When pitted dog fighting eventually fell out of favor in British coal mining areas and pubs in the early decades of the 20th century, it had little to do with laws or moralizing ideals. It had mostly to do with machine-driven increases in productivity, related improvements in worker conditions and pay, state welfare provisioning, decreasing geographical and social isolation, and the rise of a middle class that had different material concerns and affective economies. This is not to say that pitted dog fighting disappeared altogether. Rather, it stayed on in muted form in many of the regions where it started, circulating additionally into new areas of both marginalization and power, largely in secretive rural settings rather than cities, possibly in collaboration with farmers (e.g., Palmer, 1983; *Black Country Bugle*, 1971; Guttman, 1985, p. 105; Smith, 2011).

Today, the language of pitted dog fighting continues to evolve as dogmen outcross the bull and terrier with the much larger dog breeds used in ancient war and colonial conquest. This desire for larger dogs and greater bloodshed has emerged alongside investments in extreme sports, not unlike how bare-knuckle boxing and pitted dog fighting were borne together. The changing physicality of the dog-fighting lexicon highlights how heightened the survivalist stakes have become under finance capitalism, where wealth is centralized and precarity has come to striate not only the lives of the wealthy and the poor but all of planetary life.

The problem with pitted dog fighting is not that dogs fight. It is that the "pit" is made to figure as a Darwinian world *naturally* striated by survivalist competition when, in fact, it is a world made and riven by *inequalities*. In repressing and disavowing the exploitation *and* uncertainty involved in the pit's making, unconscious anxieties emerge, impelling a compensatory feedback loop of enormous masculinist proportions.

The scene of destruction therefore goes well beyond the conscious human, the dog, and the dog pit itself. It involves the unconscious, and it implicates masculinity, capitalism, and the dog–human relationship. Dog fighting lies at the nexus of a larger assemblage of practices that displace the dyadic relations of care through which the canine came to domesticate the human. What the pit shows is the degree to which masculinist abstractions have taken over, replacing relational nurturance with competition, and vulnerability with defended strength. The companionate part of the canine species has been re-routed toward obedience, the desire to please hijacked into a drama of natural competition that doesn't exist.

What is needed going forward is not a criminalizing of dog or dog fighting, but a different political economy fully invested in life. Until then, pitted dog fighting will continue alongside other exploitative forms of human and nonhuman enclosure. Perhaps the best-case scenario is that the geography of fighting be studied to explore for whom and why dog fighting continues to hold meaning. Engagement with the dog and human communities involved will not be about rescue (for example, white women's liberation of Michael Vick's dogs), which revitalizes the inequalities that made dog fighting in the first place (Nast, 2015). Rather, it will involve a more radical kind of kinship that looks to dismantle the pit at its many levels.

Acknowledgments

Many thanks to Nikki Vigneau for her assistance with this project and to Stephanie Rutherford and Shari Wilcox for their editorial insights, generosity, and scholarly encouragement. Thanks also to DePaul's University Research Council for funding copyright permissions and to Getty Images for the rights to publish the coal pit mining photograph.

Notes

1 After Abraham Darby (1678–1717) discovered that coke could be used (instead of timber) to smelt iron ore, the coal and iron-ore area of the West Midlands became a crucible for the industrial revolution. By the mid-19th century, according to *The Christian Observer*, "the face of the country lying between Wolverhampton and Birmingham, Dudley and Walsall (where coal and iron both abound), gradually assumed a beclouded aspect, which for years and years has continued to darken, until at length the district has

become known as the 'Black Country'" (1866, p. 853). Others claim that the moniker Black Country derives from the fact that large coal seams were visible on the ground's surface (BBC, 2005).
2   Coal mining's earliest division of labor was intricate and included dozens of distinct positions that machines would later replace.
3   Brown explains how he consulted with Frederic Cuvier (1773–1838) to organize his book by accepted breed types of the time. Cuvier was the younger brother of Baron George Cuvier (1769–1832) and a well-known French zoologist and comparative anatomist in his own right.
4   See Harris's entry at http://hinksoldtymebullterriers.webs.com/thestaffordshiretype.htm.
5   By the 1860s, the language of fashion and choice would be extended through standard-breed-making and dog shows, the eugenics of which combined industrial concerns with those of empire and race (Glass, 1915; Haynes, 1912, p. 712; Borrow, 1851).

References

BBC. (2005, March 15). What and where is the Black Country? Retrieved from: www.bbc. co.uk/blackcountry/content/articles/2005/03/15/where_is_the_black_country_feature. shtml.

Becke, L. (Ed.). (1899). *Old convict days*. London: T. Fisher Unwin.

Black Country Bugle. (1973). Joe Mallen: A grand old Black Country sporting personality!, 1.

Borrow, G. (1898[1851]). *Lavengro: The scholar – the gypsy – the priest*. London: John Murray.

Brown, T. (1829). Dedication, preface, introduction, and the Bull Terrier. In *Biographical sketches and authentic anecdotes of dogs, exhibiting remarkable instances of the instinct, sagacity, and social disposition of this faithful animal*. Edinburgh: Oliver & Boyd, Tweeddale-Court, and Simpkin & Marshall.

Chawner, W. (1842). *The condition and treatment of the children employed in the mines and collieries of the United Kingdom: Carefully compiled from the appendix to the first report of the [parliamentary] commissioners appointed to inquire into this subject*. London: William Strange.

Cummins, B. (2013). *Our debt to the dog*. Durham, NC: Carolina Academic Press.

Davies, L. (2011). *Mountain fighters: Lost tales of Welsh Boxing*. Cardiff, Wales: Peerless Press.

Dickey, B. (2017). *Pit bull: The battle over an American icon*. New York: Vintage Books.

Egan, P. (1821). *Real life in London*. London: Methuen and Co.

Evans, R., E. K. Gauthier, and C. J. Forsyth. (1998). Dogfighting: Symbolic expression and validation of masculinity. *Sex Roles, 39*(11/12), 825–839.

Foster, K. (2012). *I'm a good dog: Pit bulls, America's most beautiful (and misunderstood) pet*. New York: Penguin.

Glass, E. (1915). *The sporting Bull Terrier: A book of general information valuable to owners, trainers, handlers, and breeders of Bull Terriers*. Battle Creek, MI: The Dog Fancier.

Gorant, J. (2011). *The lost dogs: Michael Vick's dogs and their tale of rescue and redemption*. New York: Penguin Group (USA) Inc.

Gorant, J. (2017). *The found dogs: The fates and fortunes of Michael Vick's pitbulls, 10 years after their heroic rescue*. Bookbaby.com (vanity published).

Great Britain. Commissioners for Inquiring into the Employment and Condition of Children in Mines and Manufactories. (1842a). *Children's employment commission: First report of the commissioners. (Mines): Presented to both houses of parliament*. London: W. Clows and Sons, Stamford Street, for Her Majesty's Stationary Office.

Great Britain. Commissioners for Inquiring into the Employment and Condition of Children in Mines and Manufactories. (1842b). *Reports from the sub-commisioners and evidence from children: Appendix to first report of commissioners (Mines)*. London: W. Clows and Sons, Stamford Street, for Her Majesty's Stationary Office.

Gresley, W. S. (1883). *A glossary of terms used in coal mining*. London: E. & F. N Spon.

Griffin, E. (2005). *England's revelry: A history of popular sports and pastimes 1660–1830*. Oxford, UK: Oxford University Press.

Guttman. A. (1985). English sports spectators: The restoration to the early nineteenth century. *Journal of Sport History*, *12*(2), 103–125.

Hair, P.E.H. (1968). Mortality from violence in British coal-mines, 1800–50. *The Economic History Review*, *21*(3), 545–561.

Haynes, W. (1912, April/September). The evolution of the white 'un'. *The Outing Magazine*, *60*, 711–716.

The Illustrated London News. (1873, February 1). The strike in South Wales: A Sunday's amusement, 97.

Metcalfe, A. (1982). Organized sport in the mining communities of South Northumberland, 1800–1899. *Victorian Studies*, *25*(4), 469–495.

Nast, H. J. (2015). Pit Bulls, slavery, and whiteness in the mid- to late- nineteenth century US: Geographical trajectories; primary sources. In K. Gillespie and R. C. Collard (Eds.), *Critical animal geographies*, 127–146. New York: Routledge.

Palmer, R. (1983). The minstrel of Quarry Bank: Reminiscences of George Dunn (1887–1975). *Oral History*, *11*(1), 62–68.

Punch. (1873, February 15). Sabbatarians on strike, 66.

QR (Quarterly Review [London]). (1842). Article VI.1–3 of "The report of the commissioners for inquiring into the condition of children employed in mines, &c. with two appendices of evidence. Presented to both Houses of Parliament by command of Her Majesty. 3 vols. Folio. Pp. 2022". *70*(129), 158–195. London: John Murray.

A resident in the mining district. (1866, November). The mining district of South Staffordshire: Its condition, social and spiritual. *The Christian Observer*, 850–856.

Ritvo, H. (1986). Pride and pedigree: The evolution of the Victorian dog fancy. *Victorian Studies*, Winter, 227–253.

Rouse, P. (2015). *Sport and Ireland: A history*. New York, NY: Oxford University Press.

Shimmin, H. (1856). *Liverpool life: Its pleasures, practices, and pastimes*. Reprinted from *The Liverpool Mercury* by Liverpool: Egerton Smith and Co., Mercury Office.

Smith, R. (2011). Investigating financial aspects of dog-fighting in the UK. *Journal of Financial Crime*, *18*(4), 336–346.

"Stonehenge" [John Henry Walsh] (Ed.). (1872). The bull-terrier and the bulldog. In *The dogs of the British Islands*, 124–131. London, UK: Horace Cox.

Watkins, F. (1846). Report on schools in the Northern district. In *Minutes of the Committee of Council on Education, vol. 2*. London: Published for Her Majesty's Stationary Office, by John W. Parker.

Watson, J. (1906). *The dog book: A popular history of the dog, with practical information as to care and management of house, kennel, and exhibition dogs; and descriptions of all the important breeds, vol. 2*. New York, NY: Doubleday, Page & Company.

Watts, M. (2000). Enclosure. In C. Philo and C. Wilbert (Eds.), *Animal spaces, beastly places*, 292–304. New York: Routledge.

Waugh, E. (1855). *Sketches of Lancashire life and localities*. London: Wittaker and Co.

Youles, T., P. Fernando, T. Burton, and F. Colls. (2008). Delving in the dean: The delves: An area of unrecorded early coal mining (Part Three). *Gloucestershire Society for Industrial Archaeology Journal*, 37–52.

Part IV

The global – imperial networks and the movements of animals

12 Migration, assimilation, and invasion in the nineteenth century[1]

Harriet Ritvo

People were on the move in the nineteenth century. Millions of men and women took part in the massive transfers of human population that occurred during that period, spurred by war, famine, persecution, the search for a better life, or (most rarely) the spirit of adventure. The largest of these transfers – although by no means the only one – was from the so-called Old World to the so-called New. This is a story that has often been told, though its conclusion has been subject to repeated revision. That is to say, the consequences of these past population movements continue to unfold throughout the world, even as new movements are superimposed on them. Of course, people are not unique in their mobility, as they are not unique in most of their attributes. Other animals share our basic desires with regard to prosperity and survival, and when they move independently, they are therefore likely to have similar motives. But, like people, they don't always move independently. And, as in the human case, when the migrations of animals are controlled by others, their journeys also reveal a great deal about those who are pulling the strings. A couple of animal stories can serve as examples. They both concern creatures transported far from their native habitats by the Anglophone expansions of the nineteenth century. The motives for their original introductions a century and a half ago were rather different, as have been their subsequent fates, but they were introduced to the same widely separated shores under circumstances that resembled each other in suggestive ways.

One story concerns the English or house sparrow (*Passer domesticus*), which was apparently first introduced into the United States by a nostalgic Englishman named Nicolas Pike in 1850, and subsequently reintroduced in various locations in eastern North America. In Darwinian terms, this was the beginning of a great success story. So conspicuously did the English sparrow flourish that in 1889, the Division of Economic Ornithology and Mammalogy (part of the U.S. Department of Agriculture – an ancestor of the current Fish and Wildlife Service) devoted its first monograph to it (Barrow, 1889; Moulton *et al.*, 2010). By 1928, a Department of Agriculture survey of introduced birds made the same point by opposite means, explaining the brevity of its entry on the species on the grounds that it "receives such frequent comment that it requires no more than passing notice here" (Phillips, 1928: 49). It remains one of the commonest birds in North America, though its populations have recently suffered precipitous declines elsewhere in the world.

The sparrow's adaptation to North America may have been a triumph from the passerine point of view, but hominids soon came to a different conclusion. Although the first introduction was at mid-century, the most celebrated one occurred a decade and a half later. The *New York Times* chronicled the evolving opinions inspired by the new immigrants. In November 1868, it celebrated the "wonderfully rapid increase in the number of sparrows which were imported from England a year or so ago"; they had done "noble work" by eating the inchworms that infested the city's parks, described by the *Times* as "the intolerable plague or numberless myriads of that most disgusting shiver-producing, cold-chills-down-your-back-generating, filthy and noisome of all crawling things." The reporter praised the kindness of children who fed the sparrows and that of adults who subscribed to a fund that provided birdhouses for "young married couples"; he promised that, if they continued to thrive and devour, English sparrows would be claimed as "thoroughly naturalized citizens" (Anon, 1868: 8).

Two years later, sympathy was still strong, at least in some quarters. For example, the author of an anonymous letter to the editor of the *Times* criticized his fellow citizens in general, and Henry Bergh, the founder of the American Society for the Prevention of Cruelty to Animals, in particular, for failing to provide thirsty sparrows with water. Bergh took the allegation seriously enough to compose an immediate reply, pointing out that despite his "profound interest . . . in all that relates to the sufferings of the brute creation – great and small," neither he nor his society had authority to erect fountains in public parks (Anon, 1870a: 2; Bergh, 1870: 3). But the tide was already turning. Only a few months later, the *Times* published an article titled, "Our Sparrows. What They Were Engaged to Do and How They Have Performed Their Work. How They Increase and Multiply – Do They Starve Our Native Song-Birds, and Must We Convert Them into Pot-Pies?" (Anon, 1870b: 6).

While the English sparrow was making itself at home in New York and adjoining territories, another creature was having a very different immigrant experience far to the southwest. In the early 1850s, after the American annexation of what became Texas, California, Arizona, and New Mexico, the U.S. Army found that patrolling the vast empty territory along the Mexican frontier was a daunting task, especially in the overwhelming absence of roads. The horses and mules that normally hauled soldiers and their gear did not function efficiently in this harsh new environment. Of course, though the challenges of the desert environment were new to the U.S. Army, they were not absolutely new. The soldiers and merchants of North Africa and the Middle East had solved a similar problem centuries earlier, and some open-minded Americans were aware of this (see Bulliet, 1990).[2] Several officials serving in the dry trackless regions therefore persuaded Jefferson Davis, then the U.S. Secretary of War, that what the army needed was camels (Figure 12.1), and in 1855 Congress appropriated $30,000 to test the idea (Marsh, 1856: 210).

Acquiring camels was more expensive than acquiring sparrows, partly because they are much larger and partly because such transactions required intermediate negotiations with people, including camel owners, foreign government, and customs officials. And the animals themselves demanded significantly more attention,

Figure 12.1 Camels
Source: Goodrich (1861: 576)

which Americans familiar only with such northern ungulates as horses and cattle were ill equipped to provide. In consequence, a Syrian handler named Hadji Ali (soon anglicized to "Hi Jolly") was hired to accompany the first shipment of camels; he outlasted his charges and was ultimately buried in Quartzsite, Arizona, where his tomb, which also commemorates the original Camel Corps, now constitutes the town's primary tourist attraction.[3] A total of 75 camels survived their ocean voyages and their subsequent treks to army posts throughout the Southwest. The officers who used them on missions were, on the whole, favorably impressed, while the muleteers who took care of them tended to hold them in more measured esteem.

But these discordant evaluations did not explain the ultimate failure of the experiment. With the outbreak of the Civil War, responsibility for the camels, whose numbers had grown somewhat through natural increase, passed to the Confederacy. Even their early advocate Jefferson Davis had other priorities at that point. Some of the camels were sold to circuses, menageries, and zoos; others were simply allowed to wander away into the wild dry lands. They were sighted (and chased and hunted) with decreasing frequency during the post-war decades (Perrine, 1925). In 1901, a journalist who considered the whole episode to be "one of the comedies that may once in a while be found in even the dullest and most ponderous volumes of public records from the Government Printing Office" reported that "now and then a passenger on the Southern Pacific Railroad . . . has had a sight of some gaunt, bony and decrepit old camel . . . grown white with age, [and] become as wild and intractable as any mustang" (Griswold, 1901: 218–219).

Of course, the details of the assimilation or attempted assimilation – how many individuals were involved, whether they were wild or domesticated, where they went and where they came from, whether the enterprise succeeded or failed – made a great difference to the imported creatures as well as to the importers. Such attempts, often termed "acclimatization," became relatively frequent during the nineteenth century, though the simple desire to acclimatize was the reverse of novel. Whether so labeled or not, acclimatization has been a frequent corollary of domestication, as useful plants and animals have followed human routes of trade and migration; it thus dates from the earliest development of agriculture, 10,000 years and more ago. Indeed, much of the history of the world, at least from the perspective of environmental history, can be understood in terms of the dispersal and acclimatization of livestock and crops.

Historically and prehistorically, people have taken animals and plants along with them in order to re-establish their pastoral or agricultural way of life in a new setting. Thus the bones of domesticated animals (and the seeds and other remains of domesticated plants) can help archaeologists trace, for example, the spread of Neolithic agriculture from the various centers where it originated. (The agricultural complex that was ultimately transferred throughout the temperate world by European colonizers in the post-Columbian period, based on cattle, sheep, and goats, along with wheat, barley, peas, and lentils, was derived ultimately from the ancient farmers of the eastern Mediterranean.) Even the remains of less apparently useful (or at any rate, less edible) domesticated animals can signal human migration patterns. For example, the prevalence of orange cats in parts of northwestern Europe indicates long-ago Viking settlement, and the relative frequency (greater than further south and decreasing toward the Pacific) of robust polydactyl cats (a mutation that apparently arose in colonial Boston) along the northern range of American states indicates the westward movement of New Englanders (Todd, 1977: 100–107).

Alfred Crosby has christened the process by which this assemblage of domesticated animals and plants (along with the weeds, pests, and diseases that inevitably accompanied them) achieved their current global range, "ecological imperialism," replacing or subsuming his earlier coinage, "the Columbian exchange" (see Crosby, 1986, 1972). These labels are somewhat inconsistent in their political implications, but they both have validity. Especially with regard to plants, the Americas have transformed the rest of the world at least as much as they have been transformed by it: corn (maize) and potatoes are now everywhere. But of course

American imperialism, when it emerged, did not result from this multidirectional dissemination of indigenous vegetables. Instead it was a consequence of the final westward transfer of the combination of domesticated plants and animals initially developed in ancient southwest Asia, and gradually adapted to the colder, wetter climates of northern Europe and eastern North America.

The instigators of the wave of acclimatization attempts that crested in the late nineteenth century often claimed that their motives were similarly utilitarian. But as is often the case, their actions told a somewhat different story. The American experiences of the English sparrow and the camel suggest the much smaller scale of such transfers, though the relatively few imported sparrows ultimately populated an entire continent through their own vigorous efforts. In addition, most nineteenth-century introductions resulted from the vision or desire of a few individuals, not an entire community or society; they involved the introduction of more or less exotic animals to that community, rather than the transportation by human migrants of familiar animals along with tools and household goods in order to reestablish their economic routine. Self-conscious efforts at acclimatization also embodied assumptions and aspirations that were much more grandiose and self-confident: the notion that nature was vulnerable to human control and the desire to exercise that control by improving extant biota. In many ways acclimatization efforts seemed more like a continuation of a rather different activity, which also had ancient roots, though not quite as ancient: the keeping of exotic animals in game parks and private menageries (for the rich), and in public menageries and sideshows (for the poor). This practice similarly both reflected the wealth of human proprietors and implicitly suggested a still greater source of power, the ability to categorize and re-categorize, since caged or confined creatures – even large, dangerous ones like tigers or elephants or rhinoceroses – inevitably undermine the distinction between the domesticated and the wild.

The scale of these nineteenth-century enterprises was often paradoxical: they simultaneously displayed both hubristic grandeur in their aspirations and narrow focus and limited impact in their realizations. For example, the thirteenth Earl of Derby, whose estate at Knowsley, near Liverpool, housed the largest private collection of exotic wild animals in Britain, was one of the founders of the Zoological Society of London and served as its President from 1831 until he died in 1851. He bankrolled collecting expeditions to the remote corners of the world, and there were frequent exchanges of animals between his Knowsley menagerie and the Zoo at Regent's Park, as well as other public collections (Fisher and Jackson, 2002: 44–51). These exchanges were by no means unequal; indeed, the Earl's personal zoo was decidedly superior. At his death it covered more than 100 acres and included 318 species of birds (1272 individuals) and 94 species of mammals (345 individuals) (Fisher, 2002: 85–86). Among its denizens were bison, kangaroos, zebras, lemurs, numbats, and llamas, as well as many species of deer, antelope, and sheep. In addition to providing his animals with food, lodging, and expert veterinary attention (sometimes from the most distinguished human specialists), Derby had them immortalized by celebrated artists (including Edward Lear) when they were alive, and by expert taxidermists afterwards. But he made no plans for his menagerie, or even for any of the breeding groups it contained, to survive him. His heir, already an important politician and soon to be prime minister, had no interest in the animals and sold them at auction as soon as possible.

Late in the century, the eleventh Duke of Bedford, also a long-serving President of the Zoological Society of London (1899–1936), established a menagerie at Woburn Abbey, his Bedfordshire estate. By this time, the rationale for accumulating such a vast private collection of living animals had evolved. The Woburn park contained only ungulates (and a few other grazers, like kangaroos and wallabies): its residents included various deer, goats, cattle, gazelles, antelope, tapirs, giraffes, sheep, zebras, llamas, and asses. A summary census printed in 1905 made it clear that, unlike his distinguished predecessor, the Duke collected with a view to acclimatization. "Only those animals believed to be hardy" were selected for trial, and animals that were not "good specimens," either because of their savage dispositions or because their constitutions were not well adapted to the environment of an English park, did not survive long (Anon, 1905).

That is to say, he collected with a view to the future, hoping that his park would serve as a waystation for species that might find new homes in Britain, whether in stockyards or on public or private display. In several cases, Woburn Abbey in fact provided a refuge – or even the last refuge – for remnant populations. Before the Boxer Rebellion, the Duke secured a small herd of Père David's deer, a species otherwise exclusively maintained in the imperial parks of China (and so already extinct in the wild). An original herd of 18 had grown to 67 by 1913 (Chalmers Mitchell, 1913: 79). Since their Chinese relatives fell victim to political turmoil, all the current members of the species descend from the Woburn herd. He also nurtured the Przewalski's horse – a rare wild relative of domesticated horses and ponies, discovered (at least by European science) only in the late nineteenth century, when it was on the verge of extinction (R. L., 1901: 103).

The Duke's emphasis on preservation also echoed a shift that was to become increasingly evident in the rhetoric of zoological gardens in the course of the twentieth century. As zoo-goers will have noticed, preservation, both of individual animals and of threatened species, has loomed increasingly large in their publicity, though, of course, intention is often one thing, and results are another. Less predictive of the evolution of zoo policies was the Duke's emphasis on acclimatization. His menageries contained mostly ungulates because those are the animals that people like to eat. Although there have been occasional deviations, such as the scandal that engulfed the Atlanta Zoo in 1984, when it emerged that "a city worker was making rabbit stew and other dishes out of the surplus small animals he had bought from the zoo's children's exhibit,"[4] on the whole, modern zoos have taken care not to suggest that their charges, or the offspring of their charges, will end their days on someone's plate.

But this distinction – between natural history and agriculture, to put it one way – seemed less important in the early days of public zoos. Indeed, it hardly existed. On the contrary, the first goal mentioned in the "Prospectus" of the Zoological Society of London was to introduce new varieties of animals for "domestication or for stocking our farm-yards, woods, pleasure grounds and wastes" (Bastin,

1970: 385). To this end, along with the menagerie at Regent's Park, the young society established a breeding farm at Kingston Hill, not far to the west of London. It lasted only a few years, as the market for the stud services of zebus and zebras turned out to be small. But the notion that the zoo could supplement or enhance the British diet persisted, at least in some particularly active imaginations. Frank Buckland, an eccentric and omnivorous naturalist, successfully requested permission to cook and eat the remains of the zoo's deceased residents. Among the species he (and his unfortunate dinner guests) sampled were elephant, giraffe, and panther (that is, leopard) (Ritvo, 1987: 237–241).

Naturalists like Buckland, along with wealthy owners of private menageries, founded the Society for the Acclimatisation of Animals, Birds, Fishes, Insects and Vegetables within the United Kingdom in 1860. They were following in the footsteps of French colleagues, who had founded the Société Zoologique d'Acclimatation in 1854. But their proximate inspiration was a zoological dinner held at a London tavern in 1859, at which the gathered naturalists and menagerists enjoyed the haunch of an eland descended from the Earl of Derby's herd at Knowsley Park. The declared objects of the society were grandiose and diffuse: to introduce, acclimatize, and domesticate "all innocuous animals, birds, fishes, insects, and vegetables, whether useful or ornamental"; to perfect, propagate, and hybridize these introductions; to spread "indigenous animals, &c." within the United Kingdom; to procure "animals &c., from British Colonies and foreign countries"; and to transmit "animals, &c. from England to her colonies and foreign parts." If all these objects had been achieved, the result would have been a completely homogenized globe, at least with respect to the flora and the fauna. In fact, of course, none of them came close to realization. Despite Buckland's ambitious wish list, which included beavers and kangaroos, along with the more predictable bovids and cervids, most society members confined their attention to a scattering of birds and sheep, none of which made much impact on the resident plants and animals, whether wild or domesticated. The Society itself survived only through 1866, when it enrolled only 270 members, of which 90 were life members who had therefore lost the power of expressing disaffection; it was then absorbed by the Ornithological Society of London (Lever, 1977: 29–35; Ritvo, 1987: 239; see also Lever, 1992).

The French society was larger (2600 members in 1860, including a scattering of foreign dignitaries), longer-lasting, and more firmly grounded, both in Paris, where it controlled its own Jardin d'Acclimatation, and within a network of colonial societies (Anderson, 1992: 143–144; Osborne, 2000: 143–145). It kept elaborate records, which could be consulted by any landowner wishing to diversify his livestock. But, like those in Britain, French acclimatization efforts never had a significant local economic effect, nor did they transform the landscape. Instead, they made life a little more curious and entertaining. By the end of the century, the generalization that "animal acclimatisation in Europe is now mainly sentimental or is carried out in the interests of sport or the picturesque" applied in France as well as Britain, where, according to a commentator in the *Quarterly Review*, aficionados of the exotic could savor "the pleasure of watching [the] unfamiliar forms [of Japanese apes and American prairie dogs, as well as gazelles and zebras] amid the familiar scenery" (Anon, 1900: 199–201).

The main economic impact of French acclimatization efforts was in such warmer colonial locations as Algeria. And though the British society lacked official or quasi-official support (at least with regard to animals – Kew Gardens was at the center of a network concerned with the empire-wide distribution of plants that might produce economic benefits), the Anglophone acclimatization movement also had great (though not necessarily similar) impact outside the home islands. Acclimatization societies quickly sprang up throughout Australia and New Zealand, where members embraced a weightier mission than the one undertaken by Frank Buckland or the Duke of Bedford. They felt that new kinds of animals were not needed merely for aesthetic or culinary diversification; they were needed to repair the defects of the indigenous faunas, which lacked the "serviceable animals" found so abundantly in England, including, among others, the deer, the partridge, the rook, the hare, and the sparrow. The heavy medals struck in 1868 by the Acclimatisation Society of Victoria give a sense of the seriousness with which they approached this endeavor. One side featured a wreath of imported plants, surrounding the society's name, the other a group portrait of a hare, a swan, a goat, and an alpaca, among other desirable exotic animals.[5]

Their passion was rooted in a perception of dearth. Acclimatizers complained that while nature had provided other temperate lands with "a great profusion . . . of ruminants good for food, *not one single creature of the kind inhabits Australia!*" They were not discouraged when immigrant rabbits and sparrows began to despoil gardens and fields, merely suggesting the hair of the dog as remedy: it might be advisable to "introduce the mongoose to war against the rabbits." They continued to urge "the acclimatization of every good thing the world contains" until "the country teemed with animals introduced from other countries."[6]

As was often the case, ordinary domesticated animals were not of primary concern to the most enthusiastic and visionary acclimatizers, though in many places cattle and sheep were more influential than rabbits or rats or sparrows in converting alien landscapes into homelike ones. But, in Australia, as in Texas and Arizona, extraordinary domesticated animals could fall into another category. Similar problems – vast, trackless deserts that nevertheless required to be traversed by people and their equipment – suggested similar solutions. A few immigrant camels arrived in Australia in 1840, but the ship of the desert was not integrated into the economic life of the colony (or colonies) for several decades (see Rangan and Kull, 2009). In the 1860s, just as the Civil War deflected official interest from the American camels, their Australian conspecifics were beginning to flourish, their manifest utility outweighing the perception of some who used them, that they could be spiteful, sulky, and insubordinate (Winnecke, 1884: 1–5). They even received appreciative notice in the imperial metropolis: by 1878, *Nature* reported approvingly that they worked well when yoked in pairs like oxen, and that they remained very useful in exploring expeditions, though most labored in the service of ordinary commercial purposes (Anon, 1878: 337). They also carried materials for major infrastructure projects that brought piped water and the telegraph to the dry interior. A camel breeding stud was established in 1866; overall, in addition to homegrown animals, approximately 10,000 to 12,000 camels were imported for draft and for riding during the subsequent half century.[7] Their importance continued until the 1920s, when they were supplanted by cars and trucks – the same fate that had already befallen horses in Europe and elsewhere.

Suddenly, what had seemed an unusually successful adventure in acclimatization took on a different cast. As in the American Southwest, once the camels lost their utility, they became completely superfluous. A camel-sized pet is an expensive luxury, and there was no significant circus or zoo market for animals that had long ceased to be exotic. So some were shot and others were set free to roam by kinder-hearted owners. At this point the Australian story diverged from the American one once again. Camels had lived in Australia for at least as long as many of its human inhabitants (that is, the ones with European roots) in terms of years, and in terms of generations, they had lived there longer. They were well adapted to the harsh terrain, where they foraged and reproduced, rather than dwindling and dying. As of 2009, according to the Australian Government, their feral descendants numbered close to one million – by far the largest herd of free-living camels in the world; a year later the *Meat Trade News Daily* estimated the camel population at 1.2 million.[8] They competed for resources with other animals, wild and domesticated, and it was feared that they were disrupting fragile desert ecosystems. Like some of the elephant populations of south and southeast Asia, they were occasionally reported to terrorize small towns. After helping to build the nation, they had, it was asserted, "outstayed their welcome."[9] At least until recently, culling did not keep up with new births, and the market for camel meat that had arisen in the 1980s made even less of a dent. Unsurprisingly, in a pattern that had also emerged with regard to feral horses, burros, and pigs in North America, as officials contemplated more drastic methods that would quickly reduce the population by two-thirds, human resistance also emerged, whether based on regard for the welfare of individual camels, the hope the camels could be converted dead or alive into a profit center (meat or tourism), or the fear that large-scale eradication would require the violation of property rights.[10]

The acclimatization agenda in New Zealand was somewhat different with regard to its objects, but at least equally enthusiastic and even more persistent. Since the topography and climate of New Zealand differ greatly from those of Australia, camels were never at the top of the list of targets for introduction. But acclimatizers in both places shared the desire to convert their new homelands into the most plausible possible simulacra of their old ones. In the initial burst of enthusiasm, as elsewhere, animal introductions were scattershot – anything that appealed to individual acclimatizers. But soon the focus shifted to the re-creation half a world away of the staples of British outdoor sport: deer, game birds like pheasants and grouse, and game fish like trout and salmon. Some of these thrived, with a transformative effect on the local fauna, and others languished. The ubiquitous local societies attempted to protect them by eliminating indigenous predators. In 1906, for example, the Wellington Acclimatisation Society was taking measures to combat "the shag menace to trout."[11] In the course of the twentieth century, new perspectives on this practice emerged, and enthusiasm for acclimatization diminished – though not everywhere. The plaque on an imposing monument to trout acclimatization reads:

> This centennial plaque was presented to the Auckland Acclimatisation Society to convey the gratitude of past, present, and future generations of trout anglers in New Zealand for the society's successful importation of Californian rainbow trout ova in 1883, its hatching of the eggs in the Auckland Domain Pond and its subsequent distribution of the fish and their progeny to many New Zealand waters.[12]

In 1990, the local societies were abolished; that is to say, they were converted into fish and game councils.[13]

These examples demonstrate that utility, like many other things, is a matter of perspective. Because frivolous (or worse) as they may seem from a contemporary vantage point, the instigators of all these acclimatization attempts understood themselves to be acting in the public interest, and not just for their own idiosyncratic satisfaction.

Perhaps the most poignant demonstration of this is another well-known American saga, that of the introduction of the starling (Figure 12.2). The starting point was also New York City, the scene of the excessively successful sparrow release. In 1871, the American Acclimatization Society was founded to provide a formal institutional base for such attempts. It is widely reported, though occasionally doubted, that its moving spirit, a prosperous pharmacist named Eugene Schieffelin, wished to introduce to the United States all the birds named by Shakespeare. One reason for doubt is simply quantitative – according to a little book called *The Birds of Shakespeare*, which was published in 1916, that tally would include well over 50 species, not all of them native to Britain (Geikie, 1916). But nevertheless this notion is persistent – thus, a recent article on this topic in *Scientific American* was headlined "Shakespeare to Blame for Introduction of European Starlings to U.S." (Mirsky, 2008). Less controversially, this attempt – which also turned out to be excessively successful – was part of what the Department of Agriculture retrospectively characterized as "the many attempts to add to our bird fauna the attractive and familiar [and 'useful'] song birds of Europe" (Phillips, 1928: 48–49). The report of the 1877 annual meeting of the American Acclimatization Society, at which the starling release was triumphantly announced, also approvingly noted more or less successful releases of English skylarks, pheasants, chaffinches, and blackbirds, and looked forward to the introduction of English titmice and robins, as well as additional chaffinches, blackbirds, and skylarks – all characterized as "birds which were useful to the farmer and contributed to the beauty of the groves and fields" (Anon, 1877: 2).

The acclimatization project has often been interpreted as a somewhat naïve and crude expression of the motives that underlay nineteenth-century imperialism – intellectual and scientific, as well as political and military – more generally. This understanding is compelling but not necessarily comprehensive. There is, for one thing, a significant difference between the imposition of the European biota on the

COMMON AND BLACK STARLING (⅓ nat. size).

Figure 12.2 Starlings
Source: Lydekker (1894–1895: Vol. III: 345)

rest of the world and the transfer of exotic animals and plants to the homeland
(whether inherited or adopted). And for another, the enterprise of acclimatization
is much more likely to demonstrate the limitations of human control of nature than
the reverse – whether the targets of acclimatization shrivel and die, or whether they
reproduce with unanticipated enthusiasm. Already in the nineteenth century, introduction of exotic plants and animals could be seen as a kind of Pandora's box, at
least when they were imported into Europe or heavily Europeanized colonies or
ex-colonies. For example, to return to eastern North America, the Society for the
Protection of Native Plants (now the New England Wild Flower Society) was
founded in 1900, in order to "conserve and promote the region's native plants."[14]
It was the first such organization in the United States, but in the intervening

century, societies with similar goals have been established across the continent. The commitment to preserve native flora and fauna from the encroachment of aliens marked a turn, conscious or otherwise, from offense to defense – perhaps in the American context, to be read in conjunction with the Chinese Exclusion Act of 1882 or the more comprehensive Immigration Act of 1924. And, of course, the American context was not the only relevant one, in the nineteenth century or later; elsewhere, the defense of the native would become still more strenuous.

Notes

1 A shorter version of this chapter was originally published as 'Going Forth and Multiplying: Animal Acclimatization and Invasion', *Environmental History* 17 (2012), 404–414.

2 Bulliet (1990) gives a definitive account of the integration of camel transport into the economies and societies of the Middle East and North Africa.

3 See Woodbury, 'U.S. Camel Corps Remembered in Quartzsite, Arizona', *Out West* (2003).

4 See Schmidt, 'Civic Leaders Planning Reforms for Atlanta Zoo', *New York Times* (August 28, 1984).

5 For an image of the medal, see: http://museumvictoria.com.au/collections/items/76618/medal-acclimatisation-society-of-victoria-bronze-australia-1868.

6 Acclimatisation Society of Victoria, *First Annual Report* (1862), 8, 39 and *Sixth Annual Report* (1868), 29–30; South Australian Zoological and Acclimatization Society, *Seventh Annual Report* (1885), 7; Acclimatisation Society of Victoria, *Third Annual Report* (1864), 30 and *Fifth Annual Report* (1867), 25.

7 'Camels Australia Export'. Available at: www.camelsaust.com.au/history.htm (accessed March 30, 2011); 'A Brief History of Camels in Australia', based on *Strategies for Development* (1993) prepared by the Camel Industry Steering Committee for the Northern Territory Government. Available at: http://camelfarm.com/camels/camels_australia.html

8 'Australia: The World's Largest Camel Population', *Meat Trade News Daily*, 5 September 2010. Available at: www.meattradenewsdaily.co.uk/news/100910/australia___the_worlds_largest_camel_population_.aspx (accessed May 18, 2012); 'Camel Fact Sheet', Department of the Environment, Water, Heritage and the Arts, 2009. Available at: www.environment.gov.au/biodiversity/invasive/publications/camel-factsheet.html (accessed May 18, 2012).

9 'A Million Camels Plague Australia', *National Geographic News*, 26 October 2009. Available at: http://news.nationalgeographic.com/news/2009/10/091026-australia-camels-video-ap.html (accessed March 30, 2011).

10 'Feral Camels in Western Australia', Department of Environment and Conservation, Western Australia, 2009. Available at: www.dec.wa.gov.au/content/view/3224/1968/ (accessed March 30, 2011).

11 *The Press*, Christchurch, for 23 February 1906. Reproduced in 'Acclimatisation Societies in New Zealand'. Available at: www.pyenet.co.nz/familytrees/acclimatisation/#pye accshagmenace (accessed April 1, 2011).

12 See Tongariro River Motel, "Interesting – often forgotten – peculiarities of the Taupo trout fishery. Available at: http://www.tongariorivermotel.co.nz/catch-release-daily-bag-limit-and-other-historic-stuff/ (accessed February 11, 2018).

13 Conservation Law Reform Act 1990 031 Legislation NZ. Available at: http://legislation.knowledge-basket.co.nz/gpacts/public/text/1990/se/031se74.html (accessed April 1, 2011).

14 New England Wild Flower Society Website. Available at: www.newfs.org/about/history/?searchterm=history (accessed April 1, 2011).

References

Anderson, W. 1992. Climates of Opinion, Acclimatization in 19th-Century France and England + Victorian Botanical and Biological Thought. *Victorian Studies*, 35, 134–157.

Anon. 1868. Our Feathered Friends. *New York Times*, November 22.

Anon. 1870a. Man's Inhumanity to Birds. *New York Times*, July 22.

Anon. 1870b. Our Sparrows. *New York Times*, November 20.

Anon. 1877. American Acclimatization Society. *New York Times*, November 15.

Anon. 1878. Geographical Notes. *Nature*, 337.

Anon. 1900. New Creatures for Old Countries. *Quarterly Review*, 192, 199–200.

Anon. 1905. A Record of the Collection of Foreign Animals Kept by the Duke of Bedford in Woburn Park 1892 to 1905, copy in the British Museum (Natural History).

Barrow, W. B. 1889. *The English Sparrow (Passer Domesticus) in North America in Its Relations with Agriculture*, Washington, DC: United States Department of Agriculture Division of Economic Ornithology and Mammalogy.

Bastin, J. 1970. The First Prospectus of the Zoological Society of London: A New Light on the Society's Origins. *Journal of the Society for the Bibliography of Natural History*, 5, 385.

Bergh, H. 1870. Mr Bergh and the Sparrows: A Defence against Certain Aspersions. *New York Times*, July 23.

Bulliet, R. W. 1990. *The Camel and the Wheel*, New York: Columbia University Press.

Chalmers Mitchell, P. 1913. Zoological Gardens and the Preservation of Fauna. *Nature*, 90, 79.

Crosby, A. W. 1972. *The Columbian Exchange: Biological and Cultural Consequences of 1492*, Westport, CT: Greenwood Publishing Co.

Crosby, A. W. 1986. *Ecological Imperialism: The Biological Expansion of Europe*, Cambridge: Cambridge University Press.

Fisher, C. 2002. The Knowsley Aviary & Menagerie. In: Fisher, C. T. (ed.) *A Passion for Natural History: The Life and Legacy of the 13th Earl of Derby*, Liverpool: National Museums and Galleries of Merseyside.

Fisher, C. and Jackson, C. E. 2002. The Earl of Derby as Scientist. In: Fisher, C. T. (ed.) *A Passion for Natural History: The Life and Legacy of the 13th Earl of Derby*, Liverpool: National Museums and Galleries of Merseyside.

Geikie, A. 1916. *The Birds of Shakespeare*, Glasgow: J. Maclehouse.

Goodrich, S. G. 1861. *Illustrated History of the Animal Kingdom, Being a Systematic and Popular Description of the Habits, Structure and Classification of Animals from the Highest to the Lowest Forms, with Their Relations to Agriculture, Commerce, Manufactures, and the Arts*, New York: Derby and Jackson.

Griswold, H. T. 1901. The Camel Comedy. *Current Literature*, 31, 218–219.

Lever, C. 1977. *The Naturalized Animals of the British Isles*, London: Hutchinson.

Lever, C. 1992. *They Dined on Eland: The Story of the Acclimatisation Societies*, London: Quiller Press.

Lydekker, R. (ed.) 1894–1895. *The Royal Natural History*, Vol. 3, London: Frederick Warne.

Marsh, G. P. 1856. *The Camel: His Organization Habits and Uses Considered with Reference to His Introduction into the United States*, Boston: Gould and Lincoln.

Mirsky, S. 2008. Shakespeare to Blame for Introduction of European Starlings to U.S. *Scientific American*, May 23.

Moulton, M. P., Cropper Jr., W. P., Avery, M. L. and Moulton, L. E. 2010. The Earliest House Sparrow Introductions to North America. *Biological Invasions*, 12, 2955–2958.

Osborne, M. A. 2000. Acclimatizing the World: A History of the Paradigmatic Colonial Science. *Osiris*, 15, 135–151.

Perrine, F. S. 1925. Uncle Sam's Camel Corps. *New Mexico Historical Review*, 1, 434–444.

Phillips, J. C. 1928. *Wild Birds Introduced or Transplanted in North America*, Washington, DC: United States Department of Agriculture.

Rangan, H. and Kull, C. 2009. The Indian Ocean and the Making of Outback Australia: An Ecocultural Odyssey. In: Moorthy, S. and Jamal, A. (eds.) *Indian Ocean Studies: Cultural, Social and Political Perspectives*, New York: Routledge.

Ritvo, H. 1987. *The Animal Estate: The English and Other Creatures in the Victorian Age*, Cambridge, MA: Harvard University Press.

Ritvo, H. 2012. Going Forth and Multiplying: Animal Acclimatization and Invasion. *Environmental History*, 17, 404–414.

R. L. 1901. Przewalski's Horse at Woburn Abbey. *Nature*, 65, 103.

Schmidt, W. E. 1984. Civic Leaders Planning Reforms for Atlanta Zoo. *New York Times*, August 28. Available at: www.nytimes.com/1984/08/28/us/civic-leaders-planning-reforms-for-atlanta-zoo.html?scp=1&sq=atlanta%20zoo%20%20rabbit%20sale&st=cse

Todd, N. B. 1977. Cats and Commerce. *Scientific American*, 237, 100–107.

Winnecke, C. 1884. *Mr Winnecke's Explorations during 1883*, Adelaide: Govt Printer.

Woodbury, C. 2003. U.S. Camel Corps Remembered in Quartzsite, Arizona. *Out West*. Available at: www.outwestnewspaper.com/camels.html (accessed March 27 2011).

13 Runaways and strays

Rethinking (non)human agency in Caribbean slave societies

David Lambert

In the newspapers of Britain's Caribbean colonies from the eighteenth and early nineteenth centuries, tiny figures took flight across the pages. These fugitives in print accompanied "runaway" notices about such individuals as a "Young Negro Man named FREDRICK, belonging to Mrs. *Jane Byrne*," a "Negro Man named BUTE. He is stout and well made" and a "negro Wench named HETTY. She is stout, has full breasts, and is supposed to be at the Ridge or on board some of the ships at English Harbour" (Figure 13.1).[1] Rewards for the apprehension of these runaways were offered, as well as warnings against employing them without the owner's note of permission: one could not expect to use another's property without financial or legal consequences. Elsewhere, similar notices brought attention to other forms of property no longer in their owners' possession. Announcements of "strays," some accompanied by miniature equine or bovine fugitives, described a "DARK BAY HORSE, about 13 hands high, marked on the near buttock" and a "Brown Cow, horns sawed" (Figure 13.2).[2] Clearly, runaways and strays often went missing – but they were also found. Colonial gazettes carried notices of those "taken up" and held in workhouses and pounds side-by-side, waiting for their owners to reclaim them. Failure to do so would result in the forfeiture of the property and their sale by public auction.

The notices of runways and strays – much like advertisements in the colonial Caribbean for the sale of enslaved men, women and children, horses, mules and cattle – bear troubling similarities. For example, the official Jamaican *Gazette of Saint Jago de la Vega* from November 1782 carried notice of a missing runaway, Thomas Leishman, whose left shoulder was branded with the letters "AW," alongside one for a grey mare whose buttock was marked "ID."[3] As well as the common practice of branding, both runaways and strays also had distinguishing features that bore witness to injury and punishment: "William, a Coromantee" had a "small slit on left ear," while a "Mouse-coloured He Ass" had "two slits in each ear." There was also evidence of distant origins. "Spanish marks" upon donkeys (asses) and mules usually meant that they had been imported from Cuba, while "country marks" on the humans indicated various West African origins.

Of course, signs of scarification and tooth-filing also point to differences between runaways and strays: humans deliberately marked and altered their own bodies and those of others in culturally significant ways that nonhuman animals

> *R U N—A W A Y for eight Months past.*
>
> A Young Negro Man named FREDRICK, belonging to Mrs. *Jane Byrne*;—He has large red Eyes, works on board of Veffels, and on the Parade as a Porter. —Whoever employs him after this date without a *Note*, will make themfelves liable for the Hire allowed for a Run-away.
>
> 3d *September*, 1799.

Figure 13.1 Detail from *Antigua Journal*, 3 September 1799, p. 1

> Spanifh Town, Jan. 9, 1782.
>
> STRAYED from the GOVERNMENT PENN, about three months ago,
>
> A DARK BAY HORSE, about 13 hands high, marked on the near buttock P. E. S on top. Whoever will bring him to CHRISTIAN DYKMAN fhall receive FIVE POUNDs reward. 142

Figure 13.2 Detail from *Gazette of Saint Jago de la Vega* (Jamaica), 31 January 1782, p. 3

did not. Likewise, the human runaways could explain whose property they were – or be forced to do so – and they could also dissemble. For instance, the *St. George's Chronicle and Grenada Gazette* gave notice of a runaway who "pretends to be free, and calls himself Antoine."[4] Yet, such differences should not lead us to ignore the similarities between runaways and strays. These go beyond formal parallels in how they were represented in colonial newspapers and point to the centrality of the exercise of dominion and mastery in the Caribbean, based on hierarchical and exploitative property relations. Nor should we overlook the entangled nature of the lives of humans and nonhuman animals in colonial slave societies. To give just one example from these printed notices: "strayed" animals were sometimes seized from runaways, having (unwittingly?) aided in their flight.[5]

Focusing on the colonial Caribbean, this chapter offers an initial exploration of the *captive human–animal nexus* of which these newspaper notices are one source of evidence.[6] Rather than offer a detailed empirical discussion of the entangled nature of humans and nonhuman animals in the region's slave societies, my intention is to suggest some possible areas for research. More importantly, the chapter surveys some of the key conceptual and theoretical debates of relevance to this area. In particular, I consider the notion of "agency" that has dominated work on slavery alongside recent elaborations of this concept within the field of animal studies. Despite the vitality of the latter field (e.g. Skabelund, 2013; Few and Tortorici, 2013; Kalof, 2014; Roy and Sivasundaram, 2015), there has been little engagement with more-than-human approaches among scholars of slavery in the Americas, including the Caribbean. Indeed, this chapter seeks to encourage scholars within the field of animal studies to examine societies where human slavery existed *and* to urge historians of slavery to engage with the animal turn. While this chapter's focus is on the particular historical–geographical context of Britain's Caribbean colonies in the eighteenth and early nineteenth centuries – that is, by the time "mature" plantation societies had developed but prior to the formal abolition of human slavery – it also ranges beyond to consider, and draw on, work on slavery in the Americas more broadly.

Slavery and domestication

Drawing attention to the parallels between the status of enslaved humans and domesticated animals in the context of both Ancient and New World slavery is not new. More than twenty years ago, Karl Jacoby noted that "it appears that something about slavery as an institution frequently led to a blurring of the line that has traditionally separated human beings from domestic animals" (1994, p. 90). While we might question Jacoby's reference to tradition – what traditions? when? where? – his observation pertains to both the practices and discourses of slavery.

Many of the practices associated with the domestication of animals, such as whipping, chaining, branding and castration, have also been applied to humans as part of their enslavement. In both cases, the purpose was to enforce the master's control. Indeed, Jacoby argues that "since *homo sapiens* is a social animal, like nearly every other creature successfully domesticated by humans, one can interpret slavery as little more than the extension of domestication to humans" (1994, p. 92). The philosopher Steven Best does not hesitate to label domestication as "slavery":

> The "domestication" of animals is a euphemism for a regime of exploitation, herding, confinement, castration, forced breeding, coerced labor, hobbling, branding, ear cropping, and killing. To conquer, enslave, and claim animals as their own property, to exploit them for food, clothing, labor, transportation, and warfare, herders developed broad techniques of confinement and control, such as pens, cages, collars, chains, shackles, whips, prods, and branding irons.
>
> (Best, 2014, p. 7)

In order to elaborate conceptually the notion that slavery is "little more than the extension of domestication to humans," Jacoby draws on sociologist Orlando Patterson's characterisation of slavery as "social death" or death deferred. Famously, Patterson defined slavery as "*the permanent, violent domination of natally alienated and generally dishonoured persons*" (Patterson, 2000, p. 39; emphasis in original). That both enslaved humans and domesticated nonhuman animals are subject to "violent domination" is obvious: these are precisely the techniques of control and coercion described by Jacoby and Best, and evident in the notices of runaways and strays. But Patterson's other elements are to be found in both systems, too. Examples of "natal alienation" – the forced separation and the breaking of ties of kinship and community – include the removal of offspring from their mothers and forced transportation from the place of birth (Spiegel, 1996, pp. 45–58). Human beings' imposition of themselves in the place of the parents of infant social animals was, of course, a key aspect of domestication.

The third element of slavery involves dishonouring, what Patterson describes as the "socio-psychological" aspects of the institution. It involved the rendering of the enslaved person as worthless and without social status (Patterson, 2000, p. 37). While it may be harder to see the applicability of this to animal domestication, except perhaps in the general sense that the keeping of animals is a denial of their dignity and autonomy, the notion does encourage consideration of the more symbolic forms of denigration that characterised the captive human–animal nexus. In ideological terms, Jacoby suggests that if slavery was an institution through which human beings were treated like domestic animals, how could this be justified given that humans and livestock were not the same? The "easiest solution" for those that sought to justify human slavery, according to Jacoby, was to "invent a lesser category of humans that supposedly differed little from brute beasts." He postulates that this may have begun as an "unconscious" distinction that arose from the likelihood that societies enslaved the members of "a different linguistic group." He goes on:

> As the ability to communicate through speech is one of the most commonly made distinctions between humans and animals, the captives' lack of intelligible speech – which implied in turn a lack of rationality – most likely made them appear less than fully human. From there it was a small step to treating foreign captives like the animals they apparently resembled.
>
> (Jacoby, 1994, p. 94)

From this perspective, the practice of human dominion over animals became the basis for intra-human oppression, something that would eventually be codified in the hierarchies of "race." In the context of Caribbean slavery, this bestialisation was also manifest in such ways as the application of discourses and practices of "breeding" to enslaved humans and domesticated animals alike, as well as the etymological origins of "mulatto," referring to a person of mixed African and European parentage, which was supposedly derived from the Spanish and Portuguese words for mule (Ritvo, 1987).

The rendering of enslaved humans as akin to domestic animals is a discursive move that I have characterised elsewhere as "mastery." In turn, it comprised two sub-elements: "dominion" and "paternalism." Dominion emphasised the master's "natural" and biblically sanctioned right to own and control beasts and those deemed "sub-human." It was articulated through denigration, and whipping and branding were central to this expression of control over living property. Paternalism emphasised the master's care for captive human and nonhuman alike. This translated into self-justifying ideas that human slavery was a "civilising" institution that "rescued" enslaved people from a worse fate (in Africa) or that domesticated animals could not survive without human care. If dominion was manifest in slavery in its most brutal forms, then paternalism would come more to the fore in the development of pro-natalist and ameliorative policies, which emerged in some parts of the British Caribbean from the late eighteenth century, that were intended to reduce the reliance on the trans-Atlantic slave trade by increasing the birth rate among enslaved people (Lambert, 2015).

Of course, ideological questions could cut both ways, and the emergence of animal welfarism in the nineteenth century was closely tied to abolitionist movements. As Reinaldo Funes Monzote notes, the fact that among the supporters of animal welfare societies in England and the United States "one could find many abolitionists, also fighting to eliminate slavery and the slave trade, can be seen as part of the very same (and not entirely unproblematic) process of expanding civil rights to historically marginalized groups" (Monzote, 2013, p. 222; see also Tague, 2010). More recently, the terms of this relationship between human and nonhuman animals have been reversed by the "new abolitionism" (Best, 2014, pp. 21–49).

Jacoby postulated that the connections between human enslavement and the domestication of nonhuman animals could be explained by a "deeper connection" associated with the development of agriculture (Jacoby, 1994, p. 94). Other scholars, however, have sought to stress the differences between captive human and nonhuman labour. For example, David Brion Davis has pointed out that within human slavery the roles are – potentially – reversible: the enslaved can replace and dominate the enslaver (Davis, 2000, p. 30, footnote 10). More importantly, as Jacoby himself noted, enslaved human populations have not undergone the evolutionary transformation of neoteny, whereby juvenile traits, including passivity, have become more common among domesticated animals. Despite the longevity of particular systems of human slavery, domesticated animals have been controlled and bred for much greater periods of time – though, as noted, this has not prevented supporters of human slavery from making claims about some population groups being "natural slaves" (Jacoby, 1994). Overall, those scholars who connect the institutions of human and nonhuman animal slavery do not deny the differences, but rather insist that these are less significant than their similarities (Best, 2014, p. 32).

Jacoby's articulation of the connections between the captivity of humans and nonhuman animals is helpful but can be greatly elaborated. Mainly because of his interest in origins, much of his discussion is concerned with the pre-historic. Yet, neither domestication nor enslavement were one-off activities, nor did they remain unchanged over time. Moreover, the similarities and differences between them become particularly clear and of interest to the historian when they coincide in the same social contexts, such as the European colonies established in the Caribbean prior to human emancipation. Indeed, there is great potential for the study of the captive human–animal nexus focusing on this particular historical geography. In this context, we can consider a variety of questions. For example, how did notions of "stock" and practices of breeding change? Or ongoing efforts to "break" and "season" the servile labour force? How did the emergence of a culture of "improvement" among British metropolitan landowners in the eighteenth century affect Britain's Caribbean colonies, including the treatment of captive human and nonhuman animals?[7] Was there any reverse traffic in ideas or practices? How did mechanisation and industrialisation affect the captive human–animal nexus? After all, Caribbean plantations were "factories in the field," agro-industrial enterprises that embodied a "modernity that predated the modern" (Mintz, 1985; Scott, 2004). If the enslaved human labour of the plantations can be seen as anticipating the disciplined, routinised labour that characterised factory work, how did this impact on animal labour and ideas about the work that animals performed?[8] What ideas about the status or symbolic role of animals were brought from Africa, and what impact did they have?

In sum, the connections between human and nonhuman animal slavery that Jacoby observed suggest the value of an approach to the Caribbean colonies that addresses their more-than-human history, something that could be undertaken through recourse to the perspectives drawn from animal studies. Yet historians of slave societies in the Caribbean and elsewhere have been slow to embrace the animal turn. Why might this be?

Speciesism – the master's trope?

Animals are not wholly absent from the histories of Caribbean slave societies, but they have tended to feature as elements within local economies (e.g. Shepherd, 2009; Morgan, 1995). Meanwhile, environmental histories, which might be expected to consider relations between humans and nonhuman animals, are relatively underdeveloped in the Caribbean context, and the focus has mainly been on plants (particularly cash crops), hazards and ecological "contexts" (e.g. Schwartz, 2015; Morgan, 2015; McNeill, 2010; Richardson, 2004; Watts, 1987). This is not only because the available sources marginalise nonhumans – an issue that historians of slavery are actually well-placed to address, as I will discuss later. Rather, a major reason why scholars have been unwilling to embrace work on animals has stemmed from the dehumanising nature of slavery itself. As Lucile Desblache puts it, "because enslaved black Caribbeans were treated like beasts and were considered as 'not quite' human, there emerged a desire to establish strong boundaries between human and nonhuman animals" (Desblache, 2012, p. 125). Unsurprisingly, this has militated against the adoption of post- or more-than-human perspectives. If the animal turn has been controversial in other contexts, then it is particularly so in (former) slave societies. Even to mention animal domestication in the same breath as human slavery could be seen to be disrespectful, offensive or simply irrelevant.

If an understandable effort to inscribe a strong boundary between human and nonhuman goes someway to account for the general lack of an animal turn in studies of Caribbean slavery, then a more specific explanation relates to the foregrounding of a particular notion of "agency" within slavery studies. In a highly influential article, Walter Johnson identified "agency" as the "master trope of the New Social History" that emerged from the 1960s and that remains very influential in the historiography of the Caribbean. Tasking historians with the recovery of histories from below, this imperative has also characterised studies of slavery. A specific manifestation of this within research on slavery in the Americas was the backlash against the work of the American historian, Stanley Elkins, specifically his *Slavery: A Problem in American Institutional and Intellectual Life*. Based on what then was new research on the psychological consequences for the inmates of life in Nazi death camps, Elkins drew analogies to the "total" systems of slavery in North America to argue that antebellum slavery fostered the development of an infantilised, dependent personality type among the enslaved population that he termed the "Sambo" type (1959).[9]

Reacting against Elkins and reflective of the influence of the New Social History, as well as what Richard King terms the "transformation in black consciousness and a revival of interest in black history" in the 1960s more broadly, the task of the historian of slavery came to be seen as the effort to "give the slaves back their agency" (2001). The most common way such arguments have been framed has been in terms of demonstrating or discovering the "humanity" of enslaved people or giving them "voice." Yet, there are problems with this formulation, not least the conflation of notions of agency, humanity and resistance in ways that tend to abstract and over-simplify the lived historical experiences of enslaved people. For example, Johnson points out that agency must surely include not only acts of resistance but also of collaboration and collusion – as well as simple survival (2003, pp. 113–114). This stress on recovering humanity/agency within slavery studies has also had a chilling effect on conceptual innovations that call into question hegemonic and common-sense notions of humanity (Boster, 2013, p. 5), of which many of the approaches that characterise the animal turn are exemplary. In the context of deeply embedded ideas about the purpose of histories of slavery, a focus on nonhumans may appear to be a distraction at best and, at worst, an abandonment of a historical project that ought to be centred on recovering the humanity of the enslaved, itself seen as an extension of the fight against oppression and injustice.

There are clear reasons why scholars of (Caribbean) slavery may have been reluctant to embrace the animal turn. Yet, those who have made the "dreaded comparison" between the enslavement of humans and nonhuman animals have responded strongly to what they see as a misplaced effort to defend the boundary between human and nonhuman victims. Marjorie Spiegel is clear that "[c]omparing the suffering of animals to that of blacks (or any other oppressed group) is offensive only to the speciesist," by which she means "one who has embraced the false notion of what animals are like" (1996, p. 30; see also Whatmore, 2002, p. 32). In an echo of Audre Lorde's insistence that "the master's tools will never dismantle the master's house" (1984), Spiegel portrays speciesism as part of the "biased worldview presented by the masters." She goes on:

> To deny our similarities to animals is to deny and undermine our own power. It is to continue actively struggling to prove to our masters, past or present, that we are *similar to those who have abused us*, rather than to our fellow victims, those whom our masters have also victimized.
>
> (Spiegel, 1996, p. 30, emphasis in original)

Best puts it more bluntly, insisting that those who are "offended by efforts to make legitimate claims and analogies" between the oppressive experience of humans and nonhuman animals fail to "accept that all beings have rights," particularly "the right to be free from slavery, torture, and violent murder, and free to live an autonomous, pleasurable, peaceful existence." To reject such analogies, when made in "historically informed, factually accurate, and culturally sensitive ways" by insisting on the unique nature of human slavery, is "blatantly speciesist" (Best, 2014, p. 32). These are strong sentiments and challenging ideas, and I do not have the scope to elaborate further here. However, if we accept that the effort to establish strong boundaries between the human and nonhuman in the context of the history of slavery is problematic (though perhaps understandable), then there is both a need and an opportunity to rethink the agency of subordinated figures in the Caribbean. Indeed, it is clear from Johnson's discussion of how the typical formulation of agency serves to conflate self-directed action, humanity and resistance that the field would greatly benefit from more nuanced approaches – just as have been developed in the field of animal studies.

In a recent review, Chris Pearson identifies four approaches to the question of animal agency. The first is a straightforward denial that nonhuman animals have agency because they lack an ability to think or rationalise, and have no free will. This idea of a divide between human and animal (and nature and culture) is not universal but rather emerged in the West, with scholars variously tracing its origins to Aristotle, early Christian thought or the Renaissance. It is the basis of speciesist thought. A second approach sidesteps the question of intentionality as a requirement for agency, and instead views animals as "history-shaping agents" (2014, p. 244). While perhaps most closely associated with Actor-Network Theory, this also characterises approaches in environmental history that acknowledge how nonhumans shape the world (Latour, 2005). For example, John McNeill's *Mosquito Empire* offers "an appreciation of ecological contexts and concurrent environmental trends." He argues that the ecological changes brought about by the development of the Greater Caribbean plantation system created enhanced breeding and feeding conditions for mosquito species that transmitted yellow fever and malaria, "helping them become key actors in the geopolitical struggles of the early modern Atlantic world, if not, strictly speaking, *dramatis personae*" (McNeill, 2010, p. 3; see also Greene, 2008, p. 8). Though McNeill does not draw on Actor-Network Theory, the distinction he makes between "actors" and "*dramatis personae*" is precisely a model of animal agency that "decouples agency and intentionality."

Yet, Pearson argues that this obscures a third model of nonhuman agency wherein "animals can be agents when they act in purposeful and capable ways" (2014, pp. 244, 247). While some argue that nonhumans cannot be agents because they lack the capacity to reason, calculate and plan, not only does this objection tend to prioritise linguistically-based thought, it is also based on assumptions that current behavioural research on humans and nonhuman animals makes increasingly questionable: "While humans are starting to look less intentional and rational, animals are starting to look more so" (Pearson, 2014, p. 248).

A final perspective takes things further, understanding animal agency as "resistance," an approach that is often inspired by the New Social History, Michel de Certeau's analysis of everyday practice and James C. Scott's "weapons of the weak" (Pearson, 2014, p. 250; Scott, 1985; de Certeau, 1984). Pearson sees it as problematic to label nonhuman agency in this way because "[i]t risks projecting human motivations onto animals, thereby humanizing them." Instead, he prefers to use more neutral terms to describe how animals could "thwart" or "block" projects and schemes (2014, p. 251; see also Gillespie, 2016, pp. 122–127). My own inclination is to agree with Pearson, and I think it makes better sense to retain resistance to describe what some enslaved humans sometimes did in particular historical and geographical contexts – while also recognising that not all of what they did was resistance. This is not speciesism, but rather an attempt to maintain some terminological specificity. Indeed, I think that the second model of animal agency that Pearson presents – animals as having history-shaping capacities – offers a useful starting point which, evidence permitting, might be elaborated to consider the *purposeful* capacities of animals (Pearson's third model).

This, of course, raises issues about sources and methods. For historical scholars who tend to rely on the analysis of written or visual sources in order to assess motives or emotions, the fact that animals do not leave such evidence may appear to be an insurmountable problem. At the same time, some of the proponents of the animal turn argue that "a creative reading of primary sources, combined with insights drawn from ethology and other animal sciences' can provide insight into animals' experience, subjectivity, consciousness, and motivation" (Pearson, 2014, p. 249; see Swart, 2010, pp. 194–220). Important for my argument here is that critical and creative approaches have also been vital for the study of slave societies, because the sources left to historians are almost entirely those written by owners, managers, officials and, for later period in the British Caribbean, missionaries. In other words, working with an archive that has been shaped by captivity is a familiar methodological challenge for historians of slavery, and one they are well-placed to tackle. We are used to reading between the lines of planters' journals and letters, analysing laws and regulations for evidence of official fears, reading travellers' accounts against the grain and reconstructing everyday life under slavery through estate records. Many of the same approaches may serve to reveal the presence, effects and even purpose of animals. Likewise, such diverse and familiar sources as slave narratives, contemporary paintings and the remnants of material culture attest to the ubiquity of animals in the Caribbean landscape and the closeness of human–animal relations. Maps and surveys can be used to reconstruct the micro-historical geographies of human–animal entanglements.[10] In short, there are not merely parallels between the social status and position of enslaved humans and domesticated animals in Caribbean slave societies, but rather their co-presence serves to dramatise analogous methodological questions about exploring the historical experience and agency of dehumanised and radically marginalised beings. If conducted in a spirit of interdisciplinarity, the opportunities to learn from and experiment with methods from *across* the fields of animal studies and slavery studies are considerable.[11]

Moreover, the opportunities for rethinking the agency of subordinated figures in Caribbean slave societies go beyond the novel methods that might be suggested: there is also the issue of the substantive entanglements of domesticated animals and enslaved humans. Just as I elaborated Jacoby's general argument to suggest some of the other discursive connections between animal domestication and human slavery that deserve exploration, attention is also needed to the relationships *between* enslaved humans and captive animals in specific historical–geographical contexts. For example, unfree human and nonhuman animals laboured together at the heart of the Caribbean plantation system. Cattle provided power to drive the machinery of sugarcane processing and manure to fertilise the fields. Donkeys (asses) and mules worked to transport cut cane from field to mill, and hogsheads from estate to waterfront. Such animals had to be driven and directed, as well as fed and watered, tasks that were allotted to particular enslaved workers. They also needed to be guarded to prevent them from doing damage to crops, straying from the estate or being stolen. Certain animals also worked to maintain human enslavement: horses – elite nonhuman animals in Caribbean societies – helped the masters to intimidate enslaved humans and capture runaways (Lambert, 2015). Dogs too were "agents of control" used to terrorise and track maroons and rebels, with bloodhounds specially bred in Cuba for this purpose (Desblache, 2012, p. 125; cf. Franklin and Schweninger, 1999, pp. 160–164). Yet, nonhuman animals might also act as means of escape from slavery – a mounted runaway might get farther away, albeit also attract great suspicion if spotted. More humbly, but probably of greater significance in the long term, were the small livestock that some enslaved people were able to keep, such as chickens, pigs, or goats. Often raised so that they or their produce could be sold in Sunday markets, the money earned might ultimately contribute to manumission by self-purchase or the purchase of a family member (e.g. Pinckard, 1806, vol. 1, pp. 368–370; see also Higman, 1984, p. 207). Nor were human–animal relations merely functional. They may also provide evidence for the care of other beings and skills acquired in husbandry, handling and riding. Of course, not all relations were ones of care: animals might be injured in acts of violence by enslaved people, perhaps borne of frustration or as part of more calculated forms of 'industrial sabotage'. Animals too might bite, throw, gore, or trample such that the injured bodies of humans and nonhuman animals alike serve as records for the violent proximities of Caribbean slave societies.

Conclusions

In Jamaica in the autumn of 1816, Swain Lungren placed a notice in the *Royal Gazette* that two enslaved brothers, Charles and Swain – presumably named after his master but known as "Monkey" – had run away from his Smithfield estate in

St. George's parish in the east of the island. Accompanying them was their elderly mother, Nancy, a name perhaps evoking the African folkloric spider-trickster, Anansi, and a stolen mule. It was believed that the party had taken refuge at an animal pen, where they had "relations" (presumably human, but perhaps equine too?).[12] This vignette serves to dramatise a series of points about Caribbean slave societies that I have sought to make in this chapter, including the entanglement of human and nonhuman worlds; the bestialisation of enslaved humans; and how humans and nonhuman animals collaborated in the making – and even unmaking – of slave societies. Such vignettes provide glimpses of the Caribbean's captive human-animal nexus.

If the understandable but, ultimately, speciesist framing of human exceptionalism and agency within (Caribbean) slavery studies can be overcome, the opportunities for writing new histories of these societies are profound. We should start simply by recognising and describing the ubiquitous presence of domesticated animals in Caribbean slave societies.[13] From here, it is a matter of appreciating and elaborating the entanglements of human slavery and animal domestication, be that in terms of laws, regulations and discourses, as well as specific forms of relations that were collaborative and confrontational, caring and cruel. Furthermore, it is not simply that the nonhuman animals found in Caribbean slave societies – and their entanglements with humans – deserve greater attention, but that the theoretical and conceptual developments that have occurred under the sign of animal studies have much to offer to research on slavery *and vice versa*. With speciesism set aside, the potential for conceptual and methodological sharing and cross-fertilisation is considerable. Agency, in particular, is recast not only as a property of humans, but as something that nonhuman animals could also have, sometimes working with or against humans to maintain or undermine slavery. All of this will contribute to a more-than-human history of the Caribbean that does not downplay slavery but recognises its ubiquity for those beings that laboured in these societies.

Notes

1　See, for example, *Antigua Journal*, 4 December 1798; 3 and 24 September 1799.
2　See, for example, *Gazette of Saint Jago de la Vega* (Jamaica), 31 January 1782; *Royal Gazette* (Kingston, Jamaica), supplement, 5–12 October 1822.
3　*Gazette of Saint Jago de la Vega* (Jamaica), 21 November 1782, p. 3.
4　*St. George's Chronicle and Grenada Gazette*, 8 June 1798, p. 6.
5　*Royal Gazette* (Kingston, Jamaica), 3 August 1816, p. 24.
6　For an example of other work in this direction, see S. Seymour (2011). Mules and 'improvement': Refashioning animals and Caribbean slave plantations. Paper presented at the annual conference of the Royal Geographical Society (with the Institute of British Geographers), London, 2 September; D. Lambert (2015). Master-horse-slave: Mobility, race and power in the British West Indies, c.1780–1838. *Slavery & Abolition*, 36, 618–641.
7　The controversy surrounding the proposals for the introduction of the plough is instructive here. For an initial discussion of 'improvement' in the Caribbean context, albeit one that did not integrate non-human animals, see D. Lambert (2005). *White creole culture, politics and identity during the age of abolition*. Cambridge: Cambridge University Press, pp. 41–72.
8　See J. Clutton-Brock (1992). *Horse power: A history of the horse and the donkey in human societies*. London: Natural History Museum Publications.

9 For a discussion, see Y. Nuruddin (2003). The Sambo thesis revisited: Slavery's impact upon the African American personality. *Socialism and Democracy*, 17, 291–338.
10 See Seymour, Mules and 'improvement'.
11 On the challenges of studying agency in multidisciplinary contexts, see S. Alpern (2012). Did enslaved Africans spark South Carolina's eighteenth-century rice boom? In R. Voeks and J. Rashford (Eds.) *African ethnobotany in the Americas* (pp. 35–66). New York: Springer. This was in response to D. Eltis, P. D. Morgan, and D. Richardson (2007). Agency and diaspora in Atlantic history: Reassessing the African contribution to rice cultivation in the Americas. *American Historical Review*, 112, 1329–1358, itself a criticism of the 'black rice' hypothesis. See J. A. Carney (2001). *Black rice: The African origins of rice cultivation in the Americas*. Cambridge, MA: Harvard University Press.
12 *Royal Gazette* (Kingston, Jamaica), 14 September 1816, p. 20.
13 There has not been the scope in this chapter to discuss 'wild' and 'feral' animals, including the 'pests' that beset the plantation regime.

References

Alpern, S. (2012). Did enslaved Africans spark South Carolina's eighteenth-century rice boom? In R. Voeks and J. Rashford (Eds.) *African ethnobotany in the Americas* (pp. 35–66). New York: Springer.

Best, S. (2014). *The politics of total liberation: Revolution for the 21st century*. Basingstoke: Palgrave Macmillan.

Boster, D. H. (2013). *African American slavery and disability*. New York: Routledge.

Carney, J. A. (2001). *Black rice: The African origins of rice cultivation in the Americas*. Cambridge, MA & London: Harvard University Press.

Clutton-Brock, J. (1992). *Horse power: A history of the horse and the donkey in human societies*. London: Natural History Museum Publications.

Davis, D. B. (2000). The problem of slavery. In R. L. Paquette and L. Ferleger (Eds.) *Slavery, secession, and Southern history* (pp. 17–30). Charlottesville: University Press of Virginia.

de Certeau, M. (1984). *The practice of everyday life*. Berkeley: University of California Press.

Desblache, L. (2012). Writing relations: The crab, the lobster, the orchid, the primrose, you, me, chaos and literature. In C. Blake, C. Molloy and S. Shakespeare (Eds.) *Beyond human: From animality to transhumanism* (pp. 122–142). London: Continuum.

Elkins, S. M. (1959). *Slavery: A problem in American institutional and intellectual life*. Chicago: University of Chicago Press.

Eltis, D., Morgan, P. D. and Richardson, D. (2007). Agency and diaspora in Atlantic history: Reassessing the African contribution to rice cultivation in the Americas. *American Historical Review*, 112, 1329–1358.

Few, M. and Tortorici, Z. (Eds.) (2013). *Centering animals in Latin American history*. Durham, NC: Duke University Press.

Franklin, J. H. and Schweninger, L. (1999). *Runaway slaves: Rebels on the plantations*. Oxford: Oxford University Press.

Gillespie, K. (2016). Nonhuman animal resistance and the improprieties of live property. In I. Braverman (Ed.) *Animals, biopolitics, law: Lively legalities* (pp. 117–132). London: Routledge.

Greene, A. N. (2008). *Horses at work: Harnessing power in industrial America*. Cambridge, MA: Harvard University Press.

Higman, B. W. (1984). *Slave populations of the British Caribbean, 1807–1834*. Baltimore: Johns Hopkins University Press.

Jacoby, K. (1994). Slaves by nature? Domestic animals and human slaves. *Slavery & Abolition*, 15, 89–99.

Johnson, W. (2003). On agency. *Journal of Social History*, 37, 113–124.

Kalof, L. (2014). Introduction. In L. Kalof (Ed.) *The Oxford handbook of animal studies* (pp. 1–21). Oxford: Oxford University Press.

King, R. H. (2001). Domination and fabrication: Re-thinking Stanley Elkins' Slavery. *Slavery & Abolition*, 22, 1–28.

Lambert, D. (2005). *White creole culture, politics and identity during the age of abolition.* Cambridge: Cambridge University Press.

Lambert, D. (2015). Master-horse-slave: Mobility, race and power in the British West Indies, c.1780–1838. *Slavery & Abolition*, 36, 618–641.

Latour, B. (2005). *Reassembling the social: An introduction to actor-network theory.* Oxford: Oxford University Press.

Lorde, A. (1984). The master's tools will never dismantle the master's house. In A. Lorde (Ed.) *Sister outsider: Essays and speeches* (pp. 110–113). Trumansburg, NY: Crossing Press.

McNeill, J. R. (2010). *Mosquito empires: Ecology and war in the Greater Caribbean, 1620–1914.* Cambridge: Cambridge University Press.

Mintz, S. (1985). *Sweetness and power: The place of sugar in modern history.* New York: Viking.

Monzote, R. F. (2013). Animal labor and protection in Cuba: Changes in relationships with animals in the nineteenth century. Translated by A. Hidalgo. In M. Few and Z. Tortorici (Eds.) *Centering animals in Latin American history* (pp. 209–243). Durham, NC: Duke University Press.

Morgan, P. D. (1995). Slaves and livestock in eighteenth-century Jamaica: Vineyard Pen, 1750–1751. *The William and Mary Quarterly*, 3rd Series, 52, 47–76.

Morgan, P. D. (2015). Precocious modernity: Environmental change in the early Caribbean. In E. Sansavior and R. Scholar (Eds.) *Caribbean globalizations, 1492 to the present day* (pp. 83–104). Liverpool: Liverpool University Press.

Nuruddin, Y. (2003). The Sambo thesis revisited: Slavery's impact upon the African American personality. *Socialism and Democracy*, 17, 291–338.

Patterson, O. (2000). The constituent elements of slavery. In H. Beckles and V. Shepherd (Eds.) *Caribbean slavery in the Atlantic world* (pp. 32–41). Kingston, Jamaica: Ian Randle.

Pearson, C. (2014). History and animal agencies. In L. Kalof (Ed.) *The Oxford handbook of animal studies* (pp. 240–253). Oxford: Oxford University Press.

Pinckard, G. (1806). *Notes on the West Indies: Written during the expedition under the command of the late General Sir Ralph Abercromby; including observations on the island of Barbadoes, and the settlements captured by the British troops upon the coast of Guiana*, 3 vols. London: Printed for Longman, Hurst, Rees and Orme.

Richardson, B. C. (2004). *Igniting the Caribbean's past: Fire in British West Indian history.* Chapel Hill, NC: University of North Carolina Press.

Ritvo, H. (1987). *The animal estate: The English and other creatures in the Victorian Age.* Cambridge, MA: Harvard University Press.

Roy, R. D. and Sivasundaram, S. (Eds.) (2015). Nonhuman empires. Special section in *Comparative Studies of South Asia, Africa and the Middle East*, 35.

Schwartz, S. B. (2015). *Sea of storms: A history of hurricanes in the Greater Caribbean from Columbus to Katrina.* Princeton: Princeton University Press.

Scott, D. (2004). Modernity that predated the modern: Sidney Mintz's Caribbean. *History Workshop Journal*, 58, 191–210.

Scott, J. C. (1985). *Weapons of the weak: Everyday forms of peasant resistance*. London: Yale University Press.

Shepherd, V. (2009). *Livestock, sugar and slavery: Contested terrain in colonial Jamaica*. Kingston, Jamaica: Ian Randle.

Skabelund, A. (2013). Animals and imperialism: Recent historiographical trends. *History Compass*, 11, 801–807.

Spiegel, M. (1996). *The dreaded comparison: Human and animal slavery*, 3rd edition. New York: Mirror Books.

Swart, S. (2010). *Riding high: Horses, humans and history in South Africa*. Johannesburg: Wits University.

Tague, I. H. (2010). Companions, servants, or slaves? Considering animals in eighteenth-century Britain. *Studies in Eighteenth Century Culture*, 39, 111–130.

Watts, D. (1987). *The West Indies: Patterns of development, culture and environmental change since 1492*. Cambridge: Cambridge University Press.

Whatmore, S. (2002). *Hybrid geographies: Natures, cultures, spaces*. London: Sage.

Epilogue

14 Finding our way in the Anthropocene

Stephanie Rutherford

This book has traced the stories of animals and their human co-travelers in times past. The contributors to this volume have skillfully shown the multiplicity of ways that animals (human and otherwise) co-constitute each across space and time. While these relations may always be asymmetrical, they never operate unilaterally. As such, each of the chapters responds to Hinchliffe and Bingham's contention that "[a]ll kinds of things become more interesting once we stop assuming that 'we' are the only place to begin and end our analysis" (2008, p. 1541). In taking this provocation seriously, the authors in *Historical Animal Geographies* offer a direct response to the construction of history as the domain of the human, exploring how multispecies assemblages have always been the way in which more-than-human worlds are made.

I would also contend that the chapters presented here also have something to say about the future. Although situated in the past, the stories of multispecies collaboration and contestation offer, in some sense, a history of the present. The chapters give texture and shape to the various ways that particularly located humans have relied on nonhumans for food, labor, transportation, companionship, prestige, moral uplift, and place-making. They demonstrate both how people have attempted to use animals as ciphers, and how that effort has been frustrated by animal agency. The chapters in this volume have exposed that living is only possible through constant entanglement with the more-than-human, even if some people have worked desperately to deny this inevitable connection. What the contributors to *Historical Animal Geographies* have also shown is that while these relationships are flexible, precarious, and historically and spatially contingent, what is durable is that relationships are at the heart of all our interactions. It has not suddenly become the case that we don't need the nonhuman world to survive; rather, we have become more tightly knit together, especially in the context of what is now being termed the Anthropocene.

In order to make this argument, allow me to digress into an example which may at first seem quite distant from the subject of this book. One of the ways that humans have historically interacted with nonhumans is through violence, either intentional (via extermination) or incidental (habitat loss, land conversation, climate change). Indeed, in articulations of the Anthropocene – this articulation of the human impact on the biophysical world writ large – biodiversity loss is one of its hallmarks. This has certainly been the main thread that has run throughout my historical and present-day work on wolves and wolf hybrids (Rutherford, 2013, 2016), as well as the work of many other geographers (cf. Castree, 2015; Lorimer, 2015) and historians (cf. McNeill & Engelke, 2016; Guerrini, 2016). The notion of the "Sixth Extinction" – that we are on the precipice of a widespread extinction event similar to the elimination of the dinosaurs and 75% of life on earth in the late Cretaceous, some 65 million years ago (Kolbert, 2014) – has now become one of the indicators through which our intrusion into the biosphere can be measured. Our current extinction rate is said to be potentially 100 times the normal background rate because of a combination of habitat loss, climate change, industrialized agriculture, and overpopulation, all human-induced changes. How do we solve the problem that (some) humans are efficiently and effectively wiping all species but our own off the planet?

One solution that has been put forward by synthetic biologists is de-extinction. Named with a somewhat astonishing level of hubris as "resurrection ecology," advocates of de-extinction suggest that this could be the tool that means humans haven't damaged the planet and its nonhuman inhabitants irreparably – that we can, in fact, rewrite history and re-enliven the dead. Put another way, we might be able to erase some of the egregious sins of the Anthropocene. When people talk about de-extinction, they are talking about an array of technologies. This can be as simple as back-breeding, where selective breeding of individual animals is engineered to reproduce desired characteristics in the project of re-creation. This is being done with the auroch, the prehistoric wild version of modern cows. In de-extinction's more complex articulation, DNA from museum specimens is then "edited" into the DNA of closely related species to come up with a genetic blueprint very similar to the extinct forms (Minteer, 2015, p. 12; see also Cohen, 2014). This is not cloning, which would require a living cell, but an attempt at re-creation of a species lost: an act of creative genomics. In 2003, this notion was tested by Spanish and French biologists who raised the Pyrenean Ibex from the dead, using cells harvested in 2000 from the species' last remaining member. The animal survived for a mere 10 minutes (Zimmer, 2013). Nevertheless, this project remade the conditions of possibility for the science of de-extinction. Teams of synthetic biologists are now working on a range of efforts to re-enliven creatures, from the woolly mammoth to the Tasmanian tiger.

De-extinction has its cheerleaders. Some, like M. R. O'Connor (2015), suggest that technologies of de-extinction have become central to the ethical questions surrounding our impact on the planet, and that we need to think about possibilities of "directed evolution" (location 136 of 5430). Similarly, Stewart Brand, a long-time environmentalist and now executive director of Revive and Restore, an organization whose mission is "genetic rescue for an endangered planet," sees de-extinction as a moral imperative to "reverse the founding human mistake that inspired modern conservation" – our deeply asymmetrical relations with the nonhuman world (cited in Minteer, 2015, p. 15). In this articulation, de-extinction is a no-brainer; for those who care about the animal world, it is our ethical responsibility to re-wild the places and lifeways we have destroyed as part of the Anthropocene.

What does the example of de-extinction offer by way of conclusion to *Historical Animal Geographies*? At first glance, they seem quite distinct. While the chapters in this volume look to the past to help elaborate human–animal dynamics and practices, de-extinction seeks to re-write the most injurious of those historical relations. However, what I want to suggest is that this approach to the ethical and political dimensions thrown up by the Anthropocene flies in the face of what the careful and thoughtful analyses in this volume have shown. It is also an approach which is, in my view, surprisingly unimaginative, even as it seems the stuff of science fiction. The chapters in this book have allowed for a recasting of agency, or rather a recognition of the animal agency that was always already there. Each of the chapters has encouraged an attention to the specificities of human–nonhuman relations so as to be clear, as Erica Fudge has recently noted, that "the past is made by all its inhabitants" (2017, p. 270). By contrast, de-extinction is a technophilic fever dream that seeks to erase the intertwined histories of our multispecies world. De-extinction would seem to reify the separation between nature and culture to which some people, both in the past and today, have so closely hewn.

What the chapters in this volume have shown is that this separation is a fiction, and often a function of speciesism, but one that has been durable and pernicious. De-extinction reasserts this dualistic view, offering human mastery over nature, and does nothing to shift the narratives that got us here in the first place. Indeed, it buttresses them, because, *contra* the environmentalist mantra, extinction is not forever, but rather like a knot, undone by the application of will (and a lot of money) (Sherkow & Greely, 2013, p. 33). It represents the Anthropocene as an *opportunity* for the application of technological power. And yet, without an attention to history and geography, to power and context, the animals brought back from the dead are likely to face the same conditions which drove them to extinction to begin with. Thousands of species are close to collapse: why resurrect one that still has close living relatives in a world so vastly different, in which they would be unfamiliar? What does it mean to bring one mammoth back? At what cost? The world is still characterized by habitat loss, by climate change – by those things which drive species to extinction to begin with. But instead of dealing with these thorny and complex thickets, it obviates the tragedy and finality of death as well as our responsibility in its production. In a beautifully written piece by Thom van Dooren and Deborah Rose, they contend that mourning is necessary to a reimagined relationship between humans and animals. Drawing on philosopher Thomas Attig, they suggest that:

> Mourning is about dwelling with a loss and so coming to appreciate what it means, how the world has changed, and how we must *ourselves* change and renew our relationships if we are to move forward from here. In this context, genuine mourning should open us into an awareness of our dependence on and relationships with those countless others being driven over the edge of extinction.
>
> (van Dooren & Rose, 2013)

Like the chapters in this book, van Dooren and Rose emphasize the relationality of living in the more-human-world, starting from the notion that any future that is serious about aiming to stem the Sixth Extinction needs to understand our past to get there. De-extinction, and the way of thinking it enjoins, is a way to engineer a new kind of Anthropocene, rather to genuinely engage with the ethical obligations the complexities of entanglement offer up.

If the work on this book has taught me one thing, it is that historical animal geographies are contingent affairs. If, as Richard Rorty contends, the world, like truth, is "made rather than found," then it can be made *differently* (1989, p. 7). For me, this is a profoundly hopeful notion. Revealing and working with the historicity of human–nonhuman relations matters because it shows us that contingency and offers the possibility that things could be otherwise. Paying attention to how this has varied across space matters too, as the solutions that work for animals will be necessarily place-based and situated. And so, while these chapters are about the past, I want to suggest, by way of ending, that they are also about the futurity we want to craft. That future, Sharon and I hope, is not one which relies on human will to make life livable, but one which does justice for all the actors that make up the world, human and nonhuman alike.

References

Castree, N. (2015). The Anthropocene: A primer for geographers. *Geography*, 100(2): 66–75.

Cohen, S. (2014). The ethics of de-extinction. *Nanoethics*, 8(2): 165–178.

Fudge, E. (2017). What was it like to be a cow? In Kalof, L. (Ed.) *The Oxford handbook of animal studies* (pp. 258–278). New York: Oxford University Press.

Guerrini, A. (2016). Deep history, evolutionary history, and animals in the Anthropocene. In Keulartz, J. & Bovenkerk, B. (Eds.) *Animal ethics in the age of humans: Blurring the boundaries in human-animal relationships* (pp. 25–37). Cham, Switzerland: Springer.

Hinchliffe, S. & Bingham, N. (2008). Securing life: The emerging practices of biosecurity. *Environment and Planning A*, 40: 1534–1551.

Kolbert, E. (2014). *The sixth extinction: An unnatural history*. New York, NY: Henry Holt.

Lorimer, J. (2015). *Wildlife in the Anthropocene*. Minnesota, MN: University of Minnesota Press.

McNeill, J. & Engelke, P. (2016). *The great acceleration: An environmental history of the Anthropocene since 1945*. Cambridge, MA: Harvard University Press.

Minteer, B. A. (2015). The perils of de-extinction. *Minding Nature*, 8(1): 11–17.

O'Connor, M. R. (2015). *Resurrection science: Conservation, de-extinction and the precarious future of wild things*. New York, NY: St. Martin's Press.

Rorty, R. (1989). *Contingency, irony, and solidarity*. Cambridge: Cambridge University Press.

Rutherford, S. (2013). The biopolitical animal in Canadian and environmental studies. *Journal of Canadian Studies*, 47(3): 123–144.

Rutherford, S. (2016). A resounding success? Howling as a source of environmental history. In Thorpe, J., Rutherford, S. & Sandberg, L. A. (Eds.) *Methodological challenges in nature-culture and environmental history research* (pp. 43–54). London and New York, NY: Routledge.

Sherkow, J. S. & Greely, H. T. (2013). Genomics: What if extinction is not forever? *Science*, 340(6128): 32–33.

van Dooren, T. & Rose, D. (2013, November 2). *Keeping faith with death: Mourning and de-extinction*. Retrieved from https://thomvandooren.org/2013/11/02/keeping-faith-with-death-mourning-and-de-extinction/.

Zimmer, C. (2013, April). Bringing extinct species back to life. *National Geographic*, 233: 33–36. Retrieved from www.nationalgeographic.com/magazine/2013/04/species-revival-bringing-back-extinct-animals/.

Index

Figures are indicated by page numbers in italics and boxes are indicated by page numbers followed by *b*.